Holt Geometry

Problem Solving Workbook

HOLT, RINEHART AND WINSTON

A Harcourt Education Company

Orlando • Austin • New York • San Diego • London

Contents

Holt Geometry

Problem Solving
Understanding Points, Lines, and Planes

Use the map of part of San Antonio for Exercises 1 and 2.

1. Name a point that appears to be collinear with $\overline{EF}$. Which streets intersect at this point?

   ____

   ____

2. Explain why point A is NOT collinear with $\overline{BE}$.

   ____

   ____

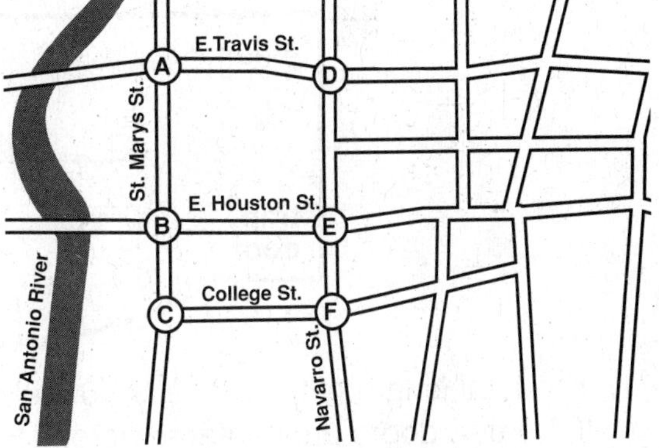

3. Suppose $\overleftrightarrow{UV}$ represents the pencil that you are using to do your homework and plane $\mathcal{P}$ represents the paper that you are writing on. Describe the relationship between $\overleftrightarrow{UV}$ and plane $\mathcal{P}$.

   ____

   ____

4. Two cyclists start at the same point, but travel along two straight streets in different directions. If they continue, how many times will their paths cross again? Explain.

   ____

   ____

Choose the best answer.

5. In a building, planes $\mathcal{W}$, $\mathcal{X}$, and $\mathcal{Y}$ represent each of the three floors; planes $\mathcal{Q}$ and $\mathcal{R}$ represent the front and back of the building; planes $\mathcal{S}$ and $\mathcal{T}$ represent the sides. Which is a true statement?

   A Planes $\mathcal{W}$ and $\mathcal{Y}$ intersect in a line.

   B Planes $\mathcal{Q}$ and $\mathcal{X}$ intersect in a line.

   C Planes $\mathcal{W}$, $\mathcal{X}$, and $\mathcal{T}$ intersect in a point.

   D Planes $\mathcal{Q}$, $\mathcal{R}$, and $\mathcal{S}$ intersect in a point.

6. Suppose point G represents a duck flying over a lake, points H and J represent two ducks swimming on the lake, and plane $\mathcal{L}$ represents the lake. Which is a true statement?

   F There are two lines through G and J.

   G The line containing G and H lies in plane $\mathcal{L}$.

   H G, H, and J are noncoplanar.

   J There is exactly one plane containing points G, H, and J.

Use the figure for Exercise 7.

7. A frame holding two pictures sits on a table. Which is NOT a true statement?

   A $\overline{PN}$ and $\overline{NM}$ lie in plane $\mathcal{T}$.

   B $\overline{PN}$ and $\overline{NM}$ intersect in a point.

   C $\overline{LM}$ and N intersect in a line.

   D P and $\overline{NM}$ are coplanar.

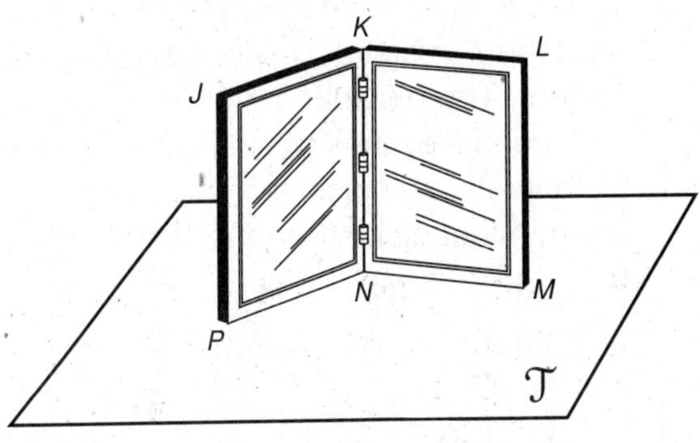

Holt Geometry

LESSON 1-2 · Problem Solving
Measuring and Constructing Segments

For Exercises 1 and 2, use the figure. It shows the top view of a stage that has three trap doors.

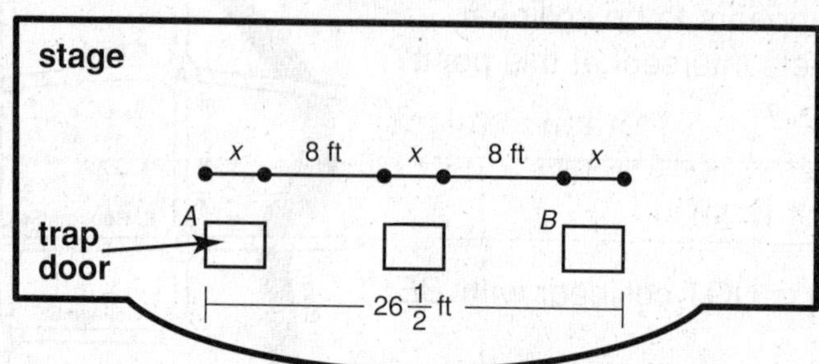

1. The total length of the stage is 76 feet. If the trap doors are centered across the stage, what is the distance from the left side of the stage to the first trap door?

2. An actor starts at point *A*, walks across the stage, and then stops at point *B* before disappearing through the trap door. How far does he walk across the stage?

3. Anna is 26 feet high on a rock-climbing wall. She descends to the 15-foot mark, rests, and then climbs down until she reaches her friend, who is 8 feet from the ground. How many feet has Anna descended?

4. Jamilla has a piece of ribbon that is 48.5 centimeters long. For her scrapbook, she cuts it into two pieces so that one piece is 4 times as long as the other. What are the lengths of the pieces?

Choose the best answer.

5. Jordan wants to adjust the shelves in his bookcase so that there is twice as much space on the bottom shelf as on the top shelf, and one and a half times more space on the middle shelf as on the top shelf. If the total height of the bookcase is 0.9 meter, how much space is the middle shelf on?

   A 0.2 m   C 0.4 m

   B 0.3 m   D 0.5 m

6. In a rowing race, the distance between the teams in first and second place is 5.7 meters. The distance between the teams in second and third place is one-third that distance. How much farther ahead is the team in first place than the team in third?

   F 7.6 m   H 2.5 m

   G 5.7 m   J 1.9 m

7. On a subway route, station C is located at the midpoint between stations A and D. Station B is located at the midpoint between stations A and C. If the distance between stations A and D is 2.4 kilometers, what is the distance between stations B and D?

   A 0.3 km   C 1.2 km

   B 0.6 km   D 1.8 km

Holt Geometry

Problem Solving
Measuring and Constructing Angles

Projection drawings are often used to represent three-dimensional molecules. The projection drawing of a methane molecule is shown below, along with the angles that are formed in the drawing.

Methane Molecule

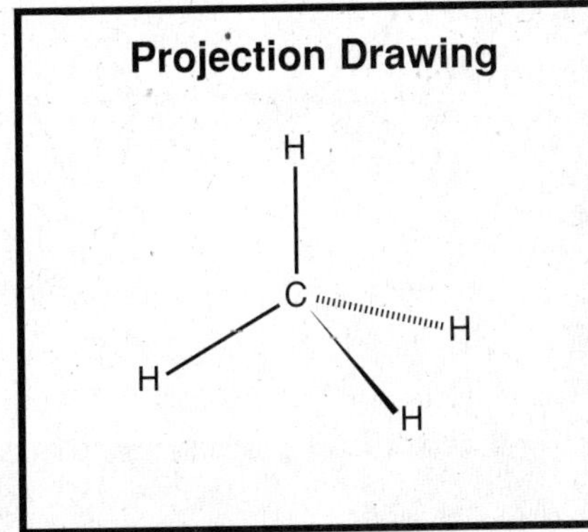

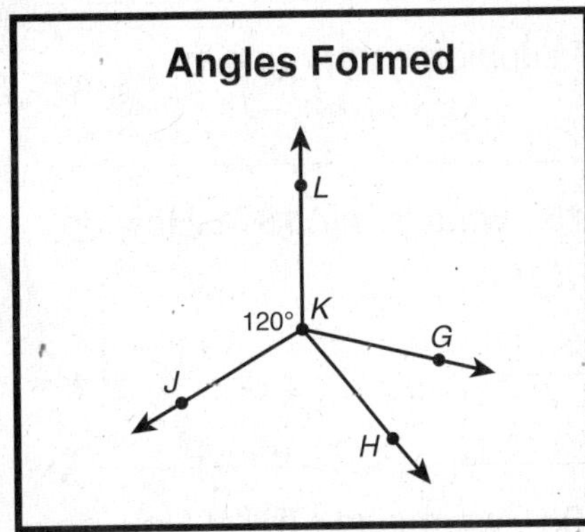

1. Name five different angles that are formed in the drawing.

   ____

2. If $m\angle LKH = m\angle JKL + 20°$ and $m\angle HKG = 37°$, what is $m\angle GKL$?   _______________

3. Find $m\angle JKH$.   _______________

The figure shows the proper way to sit at a computer to avoid straining your back or eyes. Use the figure for Exercises 4 and 5.

4. The *total viewing angle* is $\angle DAB$. If $m\angle DAC = \frac{1}{2}(m\angle ACB)$, what is the measure of the total viewing angle?

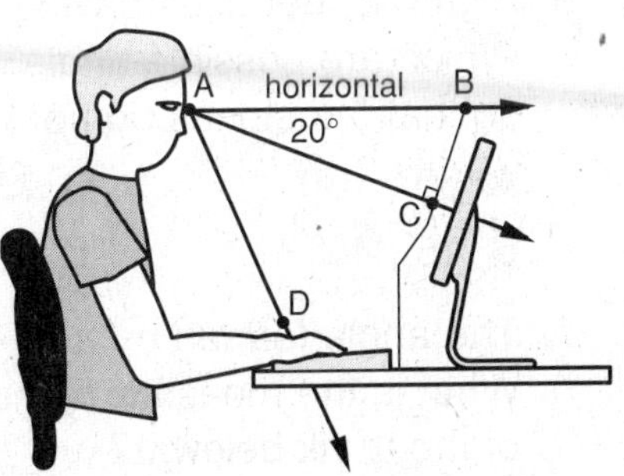

   ____

5. The *optimum viewing angle* is 38° below the horizontal. If $\overrightarrow{AV}$ is drawn to form this angle with $\overrightarrow{AB}$ and $\angle DAB$ measures 65°, what is the measure of $\angle DAV$?

   ____

Choose the best answer.

6. $\overrightarrow{QR}$ is in the interior of obtuse $\angle PQS$, and $\angle PQR$ is a right angle. Classify $\angle SQR$.

   A acute   **B** right   **C** obtuse   **D** straight

7. $\overrightarrow{VX}$ bisect $\angle WVY$, $m\angle WVX = (6x)°$, and $m\angle WVY = (16x - 42)°$. What is the value of x?

   F $\frac{21}{11}$ **G** $\frac{42}{13}$ **H** 4.2   **J** 10.5

3

Holt Geometry

LESSON 1-4
Problem Solving
Pairs of Angles

Use the drawing of part of the Eiffel Tower for Exercises 1–5.

1. Name a pair of angles that appear to be complementary.

   ____

2. Name a pair of supplementary angles.

   ____

3. If $m\angle CSW = 45°$, what is $m\angle JST$? How do you know?

   ____

   ____

4. If $m\angle FKB = 135°$, what is $m\angle BKL$? How do you know?

   ____

   ____

5. Name three angles whose measures sum to 180°.

   ____

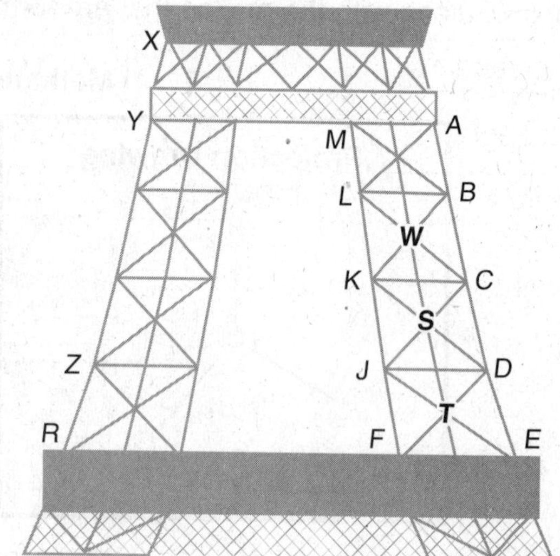

Choose the best answer.

6. A landscaper uses paving stones for a walkway. Which are possible angle measures for $a°$ and $b°$ so that the stones do not have space between them?

   A 50°, 100° **C** 75°, 105°

   B 45°, 45° **D** 90°, 80°

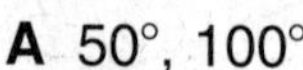

7. The angle formed by a tree branch and the part of the trunk above it is 68°. What is the measure of the angle that is formed by the branch and the part of the trunk below it?

   F 22° **H** 158°

   G 112° **J** 180°

8. $\angle R$ and $\angle S$ are complementary. If $m\angle R = (7 + 3x)°$ and $m\angle S = (2x + 13)°$, which is a true statement?

   A $\angle R$ is acute. **C** $\angle R$ and $\angle S$ are right angles.

   B $\angle R$ is obtuse. **D** $m\angle S > m\angle R$

4

Holt Geometry

LESSON 1-5 Problem Solving
Using Formulas in Geometry

Use the table for Exercises 1–6.

Figure	Perimeter or Circumference	Area
rectangle	$P = 2\ell + 2w$ or $2(\ell + w)$	$A = \ell w$
triangle	$P = a + b + c$	$A = \frac{1}{2}bh$
circle	$C = 2\pi r$	$A = \pi r^2$

Use the diagram of a hockey field for Exercises 1–4.

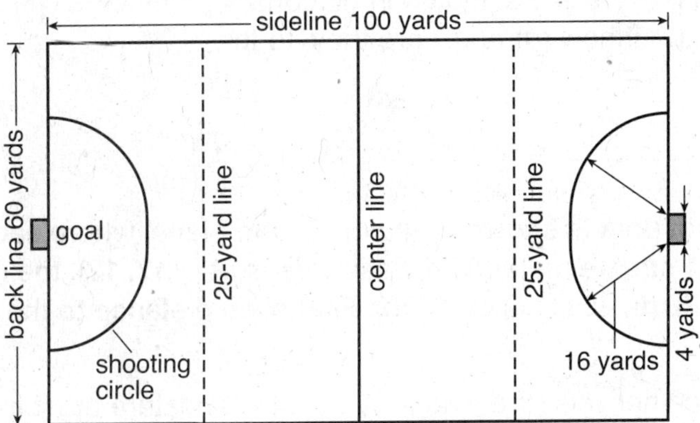

1. What is the perimeter of the field?

2. What is the area of the field?

3. What is the area of each of the shooting circles? Use 3.14 for π.

4. What is the area of the field between the two 25-yard lines?

Choose the best answer.

5. A rectangular counter 3 feet wide and 5 feet long has a circle cut out of it in order to have a sink installed. The circle has a diameter of 18 inches. What is the approximate area of the remaining countertop surface? Use 3.14 for π.

   A 13.2 ft^2 C 18.9 ft^2

   B 15.0 ft^2 D 29.5 ft^2

6. The base of a triangular garden measures 5.5 feet. Its height is 3 feet. If 4 pounds of mulch are needed to cover a square foot, how many pounds of mulch will be needed to cover the garden?

   F 8.25 lb   H 16.5 lb

   G 33 lb   J 66 lb

Holt Geometry

Problem Solving
Midpoint and Distance in the Coordinate Plane

For Exercises 1 and 2, use the diagram of a tennis court.

1. A singles tennis court is a rectangle 27 feet wide and 78 feet long. Suppose a player at corner *A* hits the ball to her opponent in the diagonally opposite corner *B*. Approximately how far does the ball travel, to the nearest tenth of a foot?

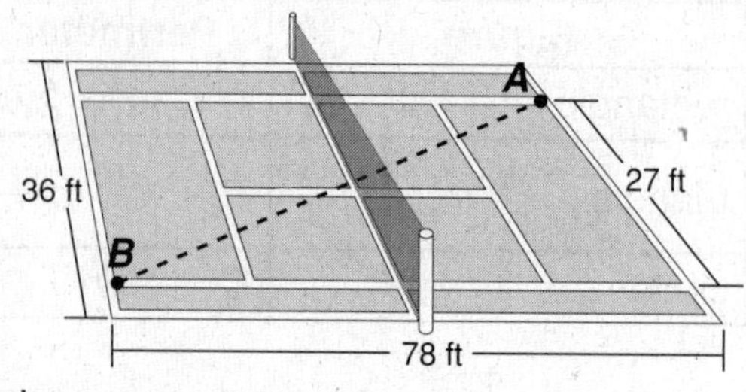

____

2. A doubles tennis court is a rectangle 36 feet wide and 78 feet long. If two players are standing in diagonally opposite corners, about how far apart are they, to the nearest tenth of a foot?

____

A map of an amusement park is shown on a coordinate plane, where each square of the grid represents 1 square meter. The water ride is at $(-17, 12)$, the roller coaster is at $(26, -8)$, and the Ferris wheel is at $(2, 20)$. Find each distance to the nearest tenth of a meter.

3. What is the distance between the water ride and the roller coaster?

____

4. A caricature artist is at the midpoint between the roller coaster and the Ferris wheel. What is the distance from the artist to the Ferris wheel?

____

Use the map of the Sacramento Zoo on a coordinate plane for Exercises 5–7. Choose the best answer.

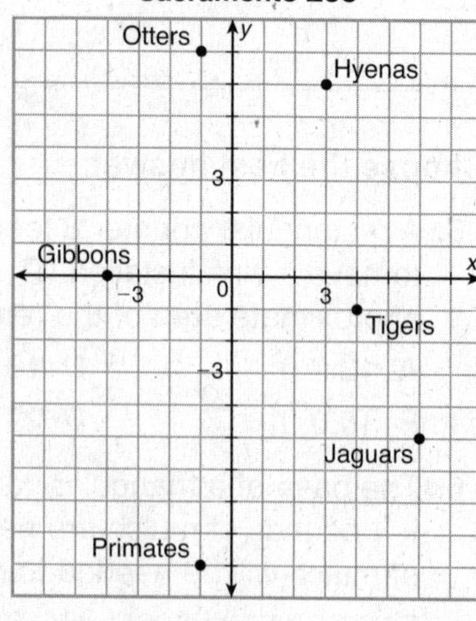

5. To the nearest tenth of a unit, how far is it from the tigers to the hyenas?

  A 5.1 units   **C** 9.9 units

  B 7.1 units   **D** 50.0 units

6. Between which of these exhibits is the distance the least?

  F tigers and primates

  G hyenas and gibbons

  H otters and gibbons

  J tigers and otters

7. Suppose you walk straight from the jaguars to the tigers and then to the otters. What is the total distance to the nearest tenth of a unit?

  A 11.4 units   **C** 13.9 units

  B 13.0 units   **D** 14.2 units

Holt Geometry

Name _________________________________ Date __________ Class __________

LESSON 1-7 **Problem Solving**
Transformations in the Coordinate Plane

Use the diagram of the starting positions of five basketball players for Exercises 1 and 2.

1. After the first step of a play, player 3 is at $(-1.5, 0)$ and player 4 is at $(1, 0.5)$. Write a rule to describe the translations of players 3 and 4 from their starting positions to their new positions.

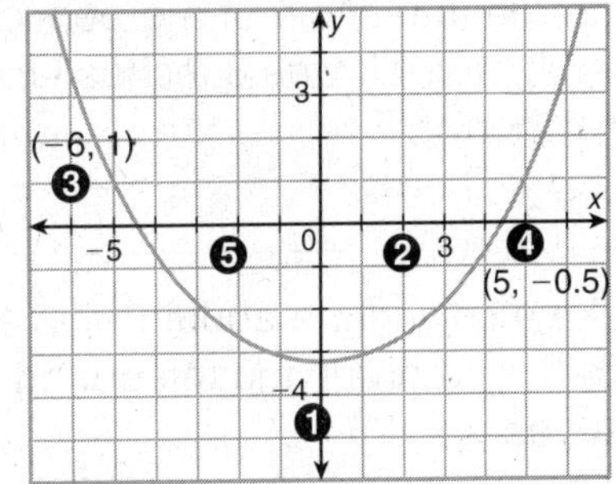

____

____

2. For the second step of the play, player 3 is to move to a position described by the rule $(x, y) \rightarrow (x - 4, y - 2)$ and player 4 is to move to a position described by the rule $(x, y) \rightarrow (x + 3, y - 2)$. What are the positions of these two players after this step of the play?

____

Use the diagram for Exercises 3–5.

3. Find the coordinates of the image of *ABCD* after it is moved 6 units left and 2 units up.

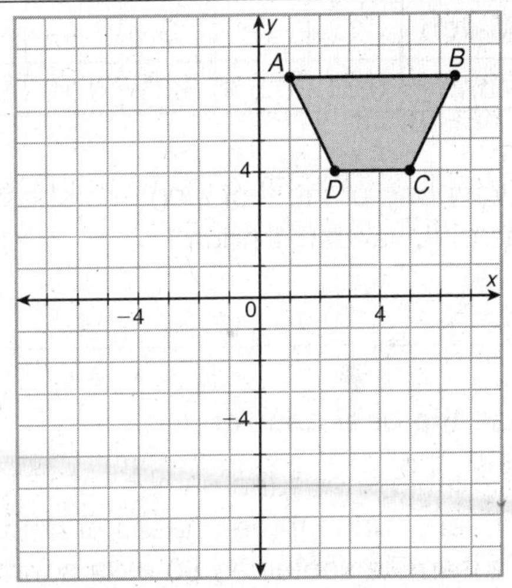

____

4. The original image is moved so that its new coordinates are $A'(-1, 7)$, $B'(-6\frac{1}{2}, 7)$, $C'(-5, 4)$, and $D'(-2\frac{1}{2}, 4)$. Identify the transformation.

____

5. The original image is translated so that the coordinates of B' are $(11\frac{1}{2}, 17)$.

What are the coordinates of the other three vertices of the image after this translation?

____

6. Triangle *HJK* has vertices $H(0, -9)$, $J(-1, -5)$, and $K(7, 8)$. What are the coordinates of the vertices after the translation $(x, y) \rightarrow (x - 1, y - 3)$?

   A $H'(-1, 12)$, $J'(-2, 8)$, $K'(6, -5)$ **C** $H'(-1, -12)$, $J'(-2, -8)$, $K'(6, 5)$

   B $H'(1, -12)$, $J'(2, -8)$, $K'(-6, 5)$ **D** $H'(1, 12)$, $J'(2, 8)$, $K'(-6, -5)$

7. A segment has endpoints at $S(2, 3)$ and $T(-2, 8)$. After a transformation, the image has endpoints at $S'(2, 3)$ and $T'(6, 8)$. Which best describes the transformation?

   F reflection across the *y*-axis   **H** rotation about the origin

   G translation $(x, y) \rightarrow (x + 8, y)$ **J** rotation about the point $(2, 3)$

Holt Geometry

LESSON 2-1 Problem Solving
Using Inductive Reasoning to Make Conjectures

The table shows the lengths of five green iguanas after birth and then after 1 year.

1. Estimate the length of a green iguana after 1 year if it was 8 inches long when it hatched.

   ____

2. Make a conjecture about the average growth of a green iguana during the first year.

Iguana	Length after Hatching (in.)	Length after 1 Year (in.)
1	10	36
2	9	34
3	11	35
4	12	35
5	10	37

The times for the first eight matches of the Santa Barbara Open women's volleyball tournament are shown. Show that each conjecture is false by finding a counterexample.

Match	1	2	3	4	5	6	7	8
Time	0:31	0:56	0:51	0:18	0:50	0:34	1:03	0:36

3. Every one of the first eight matches lasted less than 1 hour.

4. These matches were all longer than a half hour.

Choose the best answer.

5. The table shows the number of cells present during three phases of mitosis. If a sample contained 80 cells during interphase, which is the best prediction for the number of cells present during prophase?

   A 18 cells   C 40 cells

   B 24 cells   D 80 cells

Sample	Number of Cells		
	Interphase	Prophase	Metaphase
1	86	22	5
2	70	28	3
3	76	32	3
4	91	25	5
5	65	16	4
6	89	34	6

6. About 75% of the students at Jackson High School volunteer to clean up a half-mile stretch of road every year. If there are 408 students in the school this year, about how many are expected to volunteer for the clean-up?

   F 102 students   H 306 students

   G 204 students   J 333 students

7. Mara earned $25, $25, $20, and $28 in the last 4 weeks for walking her neighbor's dogs. If her earnings continue in this way, which is the best estimate for her average weekly earnings for next month?

   A $20.50   C $24.50

   B $23.33   D $25.00

8

Holt Geometry

<table>
<tr><td>**LESSON 2-2**</td></tr>
</table>

Problem Solving
Conditional Statements

1. Write the converse, inverse, and contrapositive of the conditional statement. Find the truth value of each.

If it is April, then there are 30 days in the month.

____

____

____

2. Write a conditional statement from the diagram. Then write the converse, inverse, and contrapositive. Find the truth value of each.

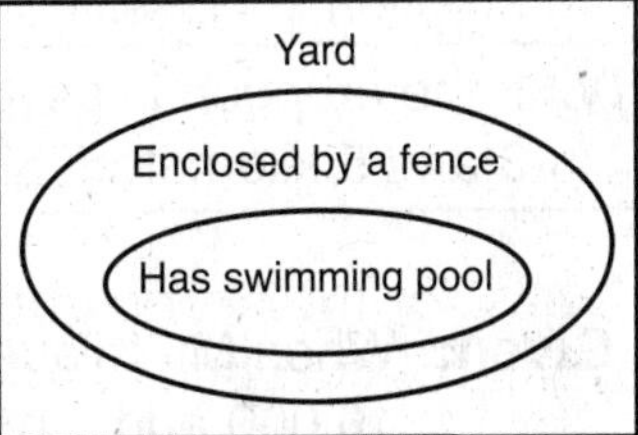

____

____

____

____

____

Use the table and the statements listed. Write each conditional and find its truth value.

p: 1777   *q*: 30 stars   *r*: after 1818   *s*: less than 50 stars

U.S. Flag	
Year	Number of Stars
1777	13
1818	20
1848	30
1959	50

3. $p \rightarrow q$ _______________________________

____

4. $r \rightarrow s$ _______________________________

5. $q \rightarrow s$ _______________________________

Choose the best answer.

6. What is the converse of "If you saw the movie, then you know how it ends"?

  A If you know how the movie ends, then you saw the movie.

  B If you did not see the movie, then you do not know how it ends.

  C If you do not know how the movie ends, then you did not see the movie.

  D If you do not know how the movie ends, then you saw the movie.

7. What is the inverse of "If you received a text message, then you have a cell phone"?

  F If you have a cell phone, then you received a text message.

  G If you do not have a cell phone, then you did not receive a text message.

  H If you did not receive a text message, then you do not have a cell phone.

  J If you received a text message, then you do not have a cell phone.

Holt Geometry

Problem Solving

LESSON 2-3

Using Deductive Reasoning to Verify Conjectures

Use the information in the table and the given statement to draw a valid conclusion for each. If a valid conclusion cannot be made, explain why not.

Volcanic Eruptions
I. A category 2 eruption produces a plume of ash 1–5 kilometers high.
II. An explosive volcano produces a volume of ash between 1 million and 10 million cubic meters.
III. If a volume of ash 10,000–1,000,000 cubic meters is produced, the eruption is classified as a category 1 eruption.
IV. If the eruption is severe, then it produces a plume of ash between 3 and 5 kilometers high.

1. **Given:** When Mt. Kilauea in Hawaii erupted, it produced a volume of ash between 10,000 and 1 million cubic meters.

____

____

2. **Given:** The eruption of a volcano in Unzen, Japan, was not explosive.

____

____

3. **Given:** The eruption of a volcano in Stromboli, Italy, was a category 2 eruption.

____

____

Choose the best answer.

4. A sports store has running shoes 25% off original prices. Andrea sees a pair of running shoes that she likes for $65.00. Which is a valid conclusion?

   A The sale price of the shoes is $40.00.

   B The sale price of the shoes is $48.75.

   C Andrea will buy the shoes.

   D Andrea will not buy the shoes.

5. If Zack makes $1\frac{1}{2}$ quarts of lemonade, then he uses 6 lemons. If Zack makes $1\frac{1}{2}$ quarts of lemonade, then he makes 4 servings. Zack uses 5 lemons. Which is a valid conclusion?

   F Zack makes 3 servings.

   G Zack makes 2 servings.

   H Zack makes 1 quart.

   J Zack does not make $1\frac{1}{2}$ quarts.

Holt Geometry

Problem Solving
Biconditional Statements and Definitions

Use the table for Exercises 1–4. Determine if a true biconditional statement can be written from each conditional. If so, then write a biconditional. If not, then explain why not.

Mountain Bike Races	Characteristics
Cross-country	A massed-start race. Riders must carry their own tools to make repairs.
Downhill	Riders start at intervals. The rider with the lowest time wins.
Freeride	Courses contain cliffs, drops, and ramps. Scoring depends on the style and the time.
Marathon	A massed-start race that covers more than 250 kilometers.

1. If a mountain bike race is mass-started, then it is a cross-country race.

2. If a mountain bike race is downhill, then time is a factor in who wins.

3. If a mountain bike race covers more than 250 kilometers, then it is a marathon race.

4. If a race course contains cliffs, drops, and ramps, then it is not a marathon race.

Choose the best answer.

5. The cat is the only species that can hold its tail vertically while it walks.

   A The converse of this statement is false.

   B The biconditional of this statement is false.

   C The biconditional of this statement is true.

   D This statement cannot be written as a biconditional.

6. Which conditional statement can be used to write a true biconditional?

   F If you travel 2 miles in 4 minutes, then distance is a function of time.

   G If the distance depends on the time, then distance is a function of time.

   H If y increases as x increases, then y is a function of x.

   J If y is not a function of x, then y does not increase as x increases.

Holt Geometry

Problem Solving
Algebraic Proof

1. Because of a recent computer glitch, an airline mistakenly sold tickets for round-trip flights at a discounted price. The equation $n(p + t) = 3298.75$ relates the number of discounted tickets sold n, the price of each ticket p, and the tax per ticket t. What was the discounted price of each ticket if 1015 tickets were sold and the tax per ticket was $1.39? Solve the equation for p. Justify each step.

2. The equation $C = 7.25s + 15.95a$ describes the total cost of admission C to the aquarium. How many student tickets were sold if the total cost for the entire class and 6 adults was $298.70? Solve the equation for s. Justify each step.

> s = number of student tickets
> a = number of adult tickets
> C = total cost of admission

Refer to the figure. Choose the best answer.

3. Which could be used to find the value of x?
   A Segment Addition Postulate
   B Angle Addition Postulate
   C Transitive Property of Congruence
   D Definition of supplementary angles

4. What is $m\angle SQR$?
   F 28°   H 61°
   G 29°   J 62°

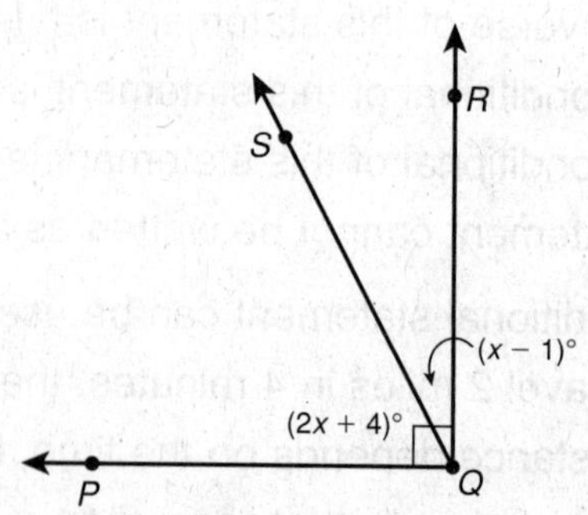

Holt Geometry

LESSON 2-6 | Problem Solving
Geometric Proof

1. Refer to the diagram of the stained-glass window and use the given plan to write a two-column proof.

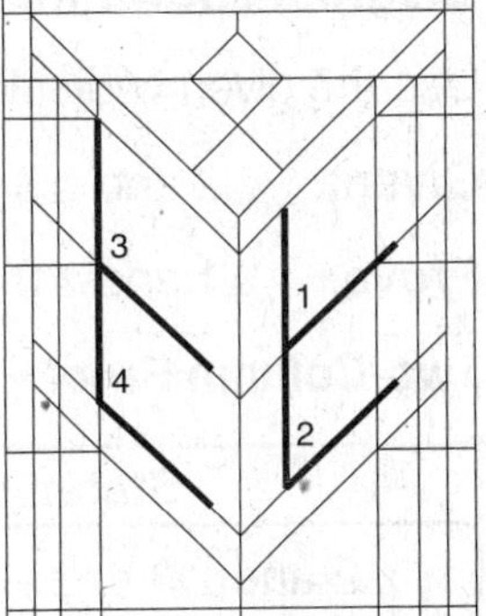

Given: ∠1 and ∠3 are supplementary.
∠2 and ∠4 are supplementary.
∠3 ≅ ∠4

Prove: ∠1 ≅ ∠2

Plan: Use the definition of supplementary angles to write the given information in terms of angle measures. Then use the Substitution Property of Equality and the Subtraction Property of Equality to conclude that ∠1 ≅ ∠2.

The position of a sprinter at the starting blocks is shown in the diagram. Which statement can be proved using the given information? Choose the best answer.

2. Given: ∠1 and ∠4 are right angles.

    A ∠3 ≅ ∠5   **C** m∠1 + m∠4 = 90°

    B ∠1 ≅ ∠4   **D** m∠3 + m∠5 = 180°

3. Given: ∠2 and ∠3 are supplementary.
∠2 and ∠5 are supplementary.

    F ∠3 ≅ ∠5   **H** ∠3 and ∠5 are complementary.

    G ∠2 ≅ ∠5   **J** ∠1 and ∠2 are supplementary.

Holt Geometry

LESSON 2-7
Problem Solving
Flowchart and Paragraph Proofs

The diagram shows the second-floor glass railing at a mall.

1. Use the given two-column proof to write a flowchart proof.

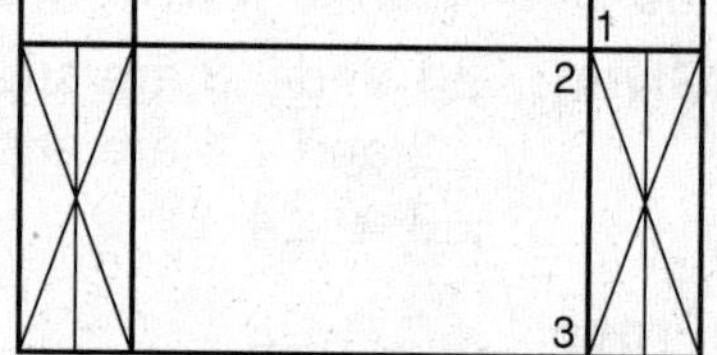

   Given: $\angle 2$ and $\angle 3$ are supplementary.

   Prove: $\angle 1$ and $\angle 3$ are supplementary.

Two-Column Proof:

Statements	Reasons
1. $\angle 2$ and $\angle 3$ are supplementary.	1. Given
2. $m\angle 2 + m\angle 3 = 180°$	2. Def. of supp. ∠s
3. $\angle 2 \cong \angle 1$	3. Vert. ∠s Thm.
4. $m\angle 2 = m\angle 1$	4. Def. of $\cong$ ∠s
5. $m\angle 1 + m\angle 3 = 180°$	5. Subst.
6. $\angle 1$ and $\angle 3$ are supplementary.	6. Def. of supp. ∠s

Choose the best answer.

2. Which would NOT be included in a paragraph proof of the two-column proof above?

   A Since $\angle 2$ and $\angle 3$ are supplementary, $m\angle 2 = m\angle 3$.

   B $\angle 2 \cong \angle 1$ by the Vertical Angles Theorem.

   C Using substitution, $m\angle 1 + m\angle 3 = 180°$.

   D $m\angle 2 = m\angle 1$ by the definition of congruent angles.

Holt Geometry

Problem Solving
Lines and Angles

**Use the diagram of the rectangular box for Exercises
1 and 2. Refer to the diagram to help justify your answer.**

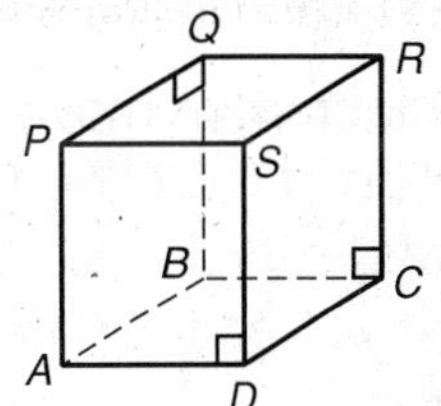

1. Is the relationship "is skew to" transitive?

____

____

2. If a segment is skew to one of two parallel segments, must it be skew to the other?

____

Use the flag of Puerto Rico for Exercises 3 and 4.

3. If ∠DFC and ∠ACF are same-side interior angles,
   identify the transversal.

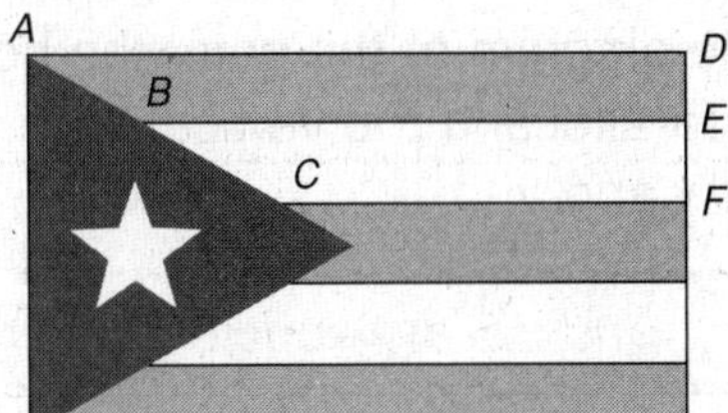

____

4. Name a pair of alternate interior angles if the
   transversal is $\overline{BE}$.

____

Choose the best answer.

5. Describe the type of lines suggested by the two skis of a person water skiing.

   A intersecting lines

   B parallel lines

   C perpendicular lines

   D skew lines

6. Describe the type of lines suggested by the paths of two people at a fair when
   one person is riding the aerial ride from one end of the fair to the other, and
   the other person is walking in a different direction on the ground.

   F intersecting   **H** perpendicular

   G parallel   **J** skew

7. In the quilt pattern, which is a true statement
   about the angles formed by the transversal
   $\overline{HK}$ and $\overline{HM}$ and $\overline{JL}$?

   A ∠LSK and ∠PHQ are corresponding angles.

   B ∠JSQ and ∠JQH are corresponding angles.

   C ∠LSK and ∠QSJ are same-side interior angles.

   D ∠PHQ and ∠RLS are same-side interior angles.

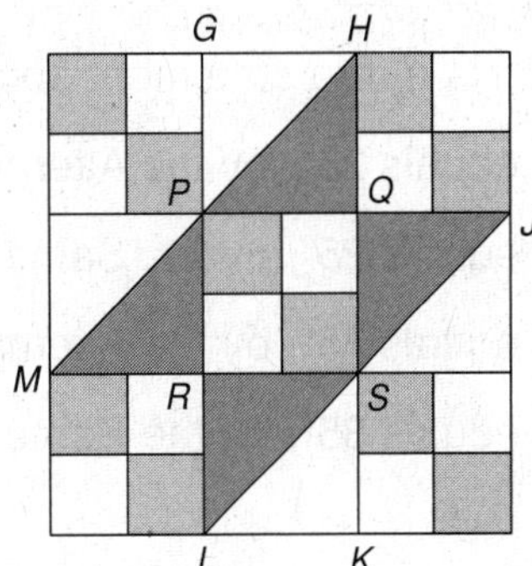

Holt Geometry

Problem Solving
Angles Formed by Parallel Lines and Transversals

Find each value. Name the postulate or theorem that you used to find the values.

1. In the diagram of movie theater seats, the incline of the floor, *f*, is parallel to the seats, *s*.

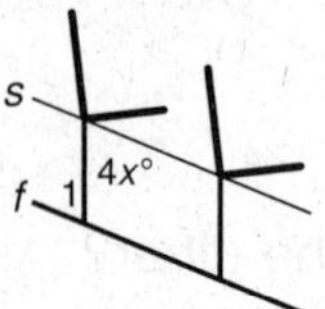

   If $m\angle 1 = 68°$, what is *x*?

   ____

2. In the diagram, roads *a* and *b* are parallel.

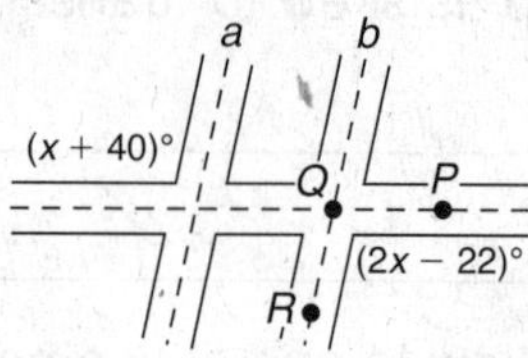

   What is the measure of $\angle PQR$?

   ____

3. In the diagram of the gate, the horizontal bars are parallel and the vertical bars are parallel. Find *x* and *y*.

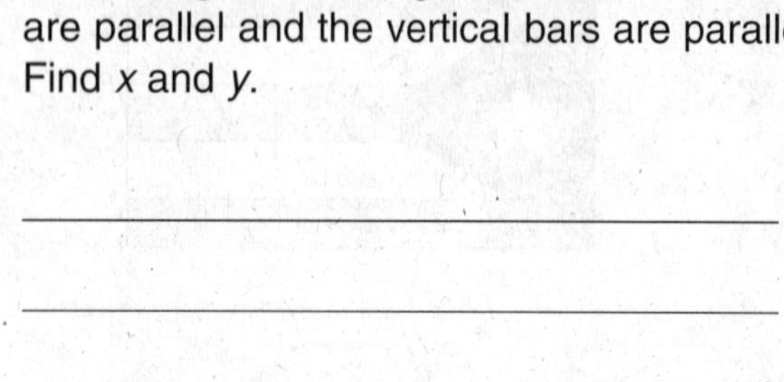

   ____

   ____

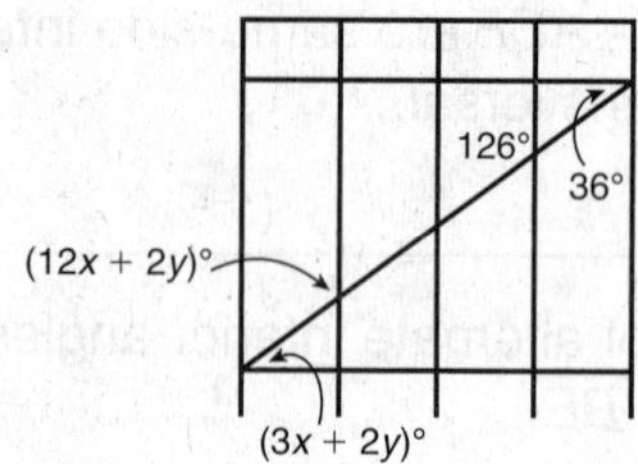

Use the diagram of a staircase railing for Exercises 4 and 5. $\overline{AG} \parallel \overline{CJ}$ and $\overline{AD} \parallel \overline{FJ}$. **Choose the best answer.**

4. Which is a true statement about the measure of $\angle DCJ$?

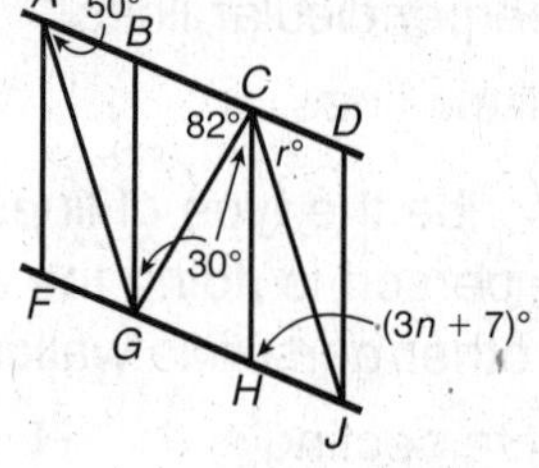

   A It equals 30°, by the Alternate Interior Angles Theorem.

   B It equals 30°, by the Corresponding Angles Postulate.

   C It equals 50°, by the Alternate Interior Angles Theorem.

   D It equals 50°, by the Corresponding Angles Postulate.

5. Which is a true statement about the value of *n*?

   F It equals 25°, by the Alternate Interior Angles Theorem.

   G It equals 25°, by the Same-Side Interior Angles Theorem.

   H It equals 35°, by the Alternate Interior Angles Theorem.

   J It equals 35°, by the Same-Side Interior Angles Theorem.

Holt Geometry

Problem Solving
Proving Lines Parallel

1. A bedroom has sloping ceilings as shown. Marcel is hanging a shelf below a rafter. If $m\angle 1 = (8x - 1)°$, $m\angle 2 = (6x + 7)°$, and $x = 4$, show that the shelf is parallel to the rafter above it.

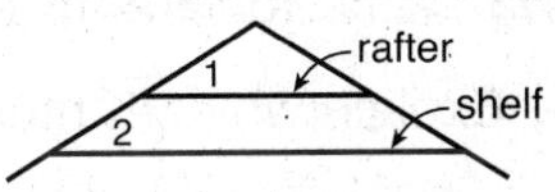

2. In the sign, $m\angle 3 = (3y + 7)°$, $m\angle 4 = (5y + 5)°$, and $y = 21$. Show that the sign posts are parallel.

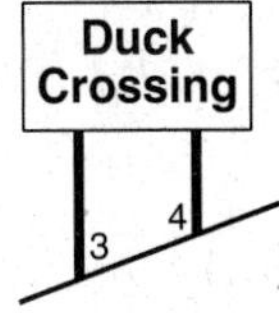

Choose the best answer.

3. In the bench, $m\angle EFG = (4n + 16)°$, $m\angle FJL = (3n + 40)°$, $m\angle GKL = (3n + 22)°$, and $n = 24$. Which is a true statement?

   A $\overline{FG} \parallel \overline{HK}$ by the Converse of the Corr. ∠s Post.

   B $\overline{FG} \parallel \overline{HK}$ by the Converse of the Alt. Int. ∠s Thm.

   C $\overline{EJ} \parallel \overline{GK}$ by the Converse of the Corr. ∠s Post.

   D $\overline{EJ} \parallel \overline{GK}$ by the Converse of the Alt. Int. ∠s Thm.

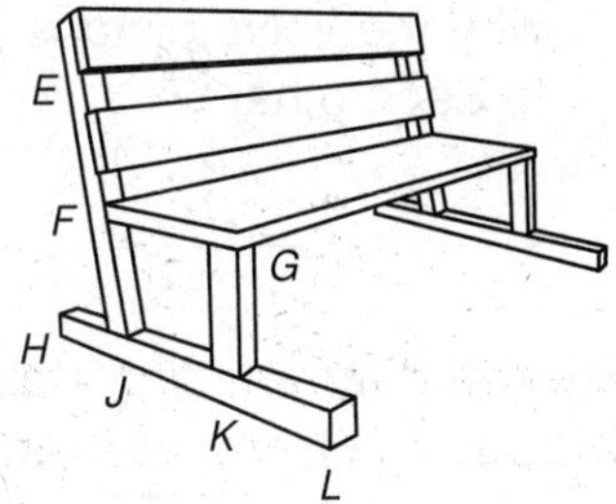

4. In the windsurfing sail, $m\angle 5 = (7c + 1)°$, $m\angle 6 = (9c - 1)°$, $m\angle 7 = 17c°$, and $c = 6$. Which is a true statement?

   F $\overline{RV}$ is parallel to $\overline{SW}$.

   G $\overline{SW}$ is parallel to $\overline{TX}$.

   H $\overline{RT}$ is parallel to $\overline{VX}$.

   J Cannot conclude that two segments are parallel

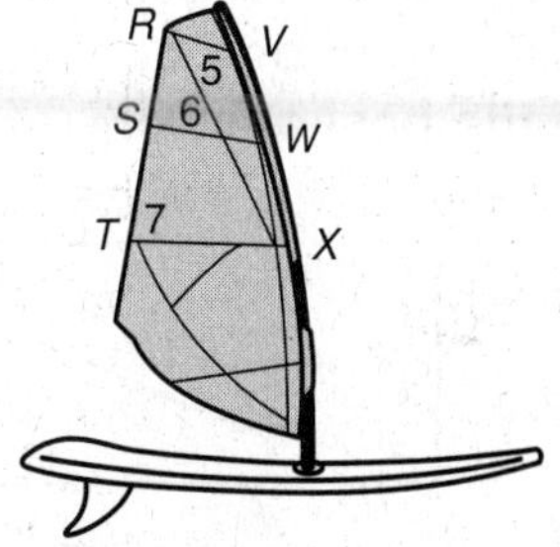

The figure shows Natalia's initials, which are monogrammed on her duffel bag. Use the figure for Exercises 5 and 6.

5. If $m\angle 1 = (4x - 24)°$, $m\angle 2 = (2x + 8)°$, and $x = 16$, show that the sides of the letter N are parallel.

6. If $m\angle 3 = (7x + 13)°$, $m\angle 4 = (5x + 35)°$, and $x = 11$, show that the sides of the letter H are parallel.

Holt Geometry

<table><tr><td>**LESSON**
3-4</td><td></td></tr></table>

Problem Solving
Perpendicular Lines

A wall rack for holding CDs is shown. Use the figure for Exercises 1 and 2.

1. Explain why $\overline{HK}$ must be perpendicular to $\overline{KL}$.

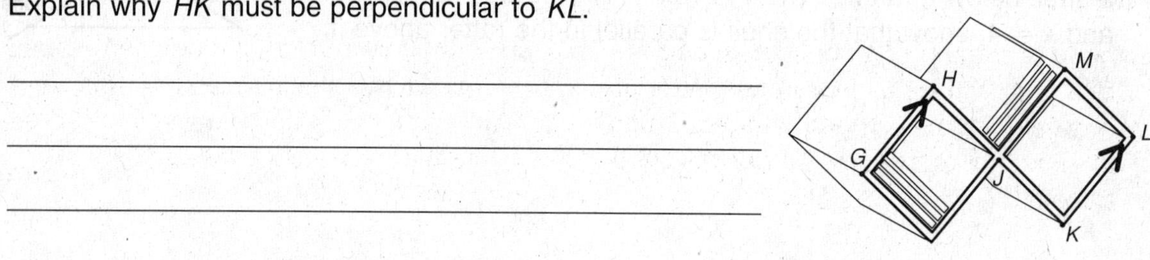

____

____

____

____

2. If $\overline{JM} \perp \overline{HK}$, explain why $\overline{JM} \parallel \overline{GH}$.

____

____

____

3. The valve pistons on a trumpet are all perpendicular to the lead pipe. Explain why the valve pistons must be parallel to each other.

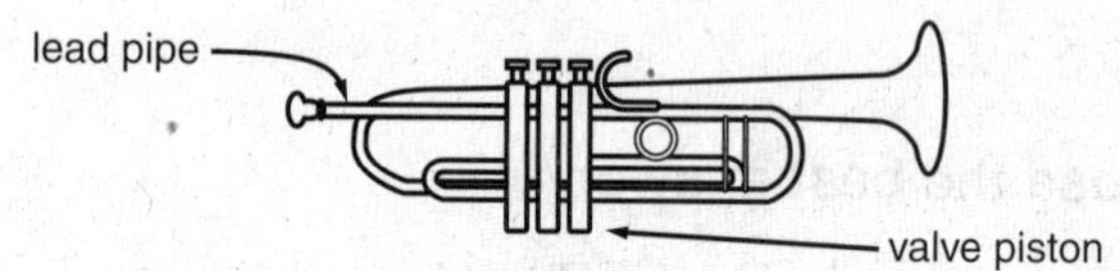

____

____

____

**Use the diagram of a bocce court for Exercises 4 and 5.
Choose the best answer.**

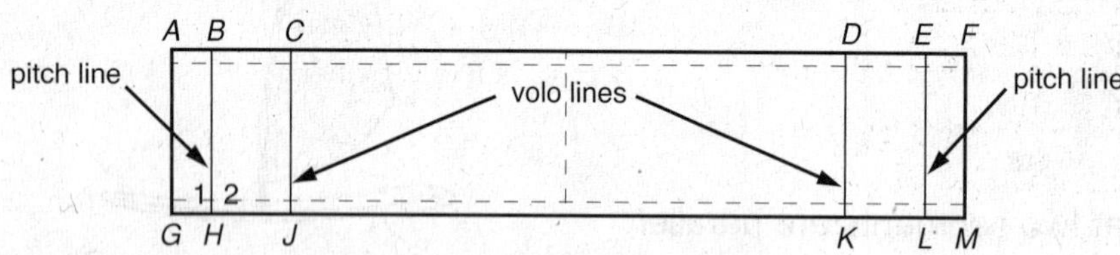

4. If $m\angle 1 = m\angle 2$, what can you conclude?

  A $\overline{BH} \perp \overline{GJ}$ **C** $\overline{BH} \parallel \overline{CJ}$

  B $\overline{AC} \perp \overline{BH}$ **D** $\overline{AC} \parallel \overline{GJ}$

5. The pitch lines are parallel, and the first pitch line is perpendicular to the long sides of the court. Which is a correct conclusion?

  F $\overline{BH} = \overline{CJ}$ **H** $\overline{EL} \perp \overline{AF}$

  G $\overline{BH} \parallel \overline{CJ}$ **J** $\overline{DK} \perp \overline{AF}$

Holt Geometry

<table><tr><td>LESSON
3-5</td><td>

Problem Solving
Slopes of Lines
</td></tr></table>

Graph the line that represents each situation. Then find and interpret the slope of the line.

1. Mara is jogging at a constant speed. She jogs 2 miles in 14 minutes. After 35 minutes, she has jogged 5 miles. Graph the line that represents Mara's distance traveled.

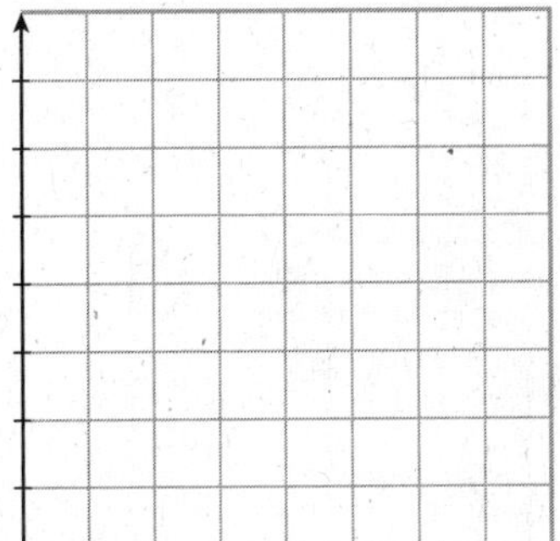

2. A turtle swimming at a constant speed travels 12 miles by 3:00 P.M. and 28 miles by 7:00 P.M. Graph the line that represents the turtle's distance traveled.

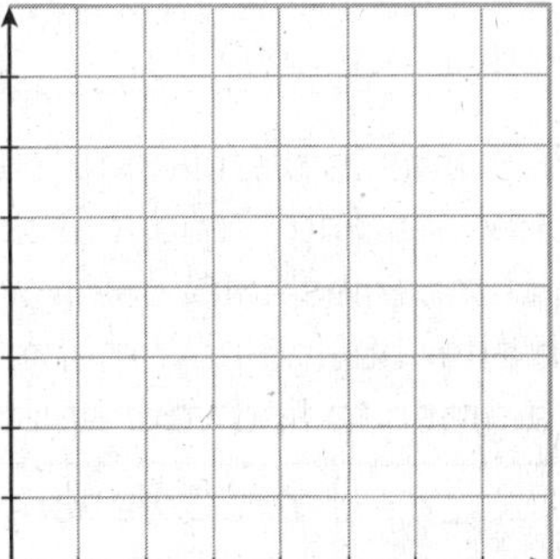

Choose the best answer.

3. A hang glider who started at 7:55 A.M. has traveled at a constant speed as shown in the table.

Time	Distance Traveled
8:00 A.M.	2 mi
8:30 A.M.	14 mi

If the line that represents the hang glider's distance traveled is graphed, which is a true interpretation of the slope?

A The hang glider is traveling at an average speed of 24 miles per hour.

B The hang glider is traveling at an average speed of 16 miles per hour.

C The hang glider is traveling at an average speed of 12 miles per minute.

D The hang glider is traveling at an average speed of 7 miles per minute.

4. The line represents the distance traveled by an in-line skater traveling at a constant speed. What is the rate of change represented in the graph?

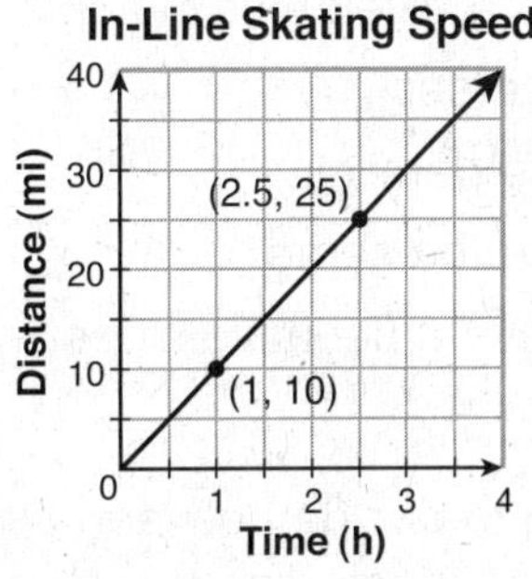

F 25 mi/h

G 15 mi/h

H 10 mi/h

J 0.1 mi/h

19

Holt Geometry

<table><tr><td>LESSON
3-6</td><td></td></tr></table>

Problem Solving
Lines in the Coordinate Plane

Use the following information for Exercises 1 and 2. Josh can order 1 color ink cartridge and 2 black ink cartridges for his printer for $78. He can also order 1 color ink cartridge and 1 black ink cartridge for $53.

1. Let x equal the cost of a color ink cartridge and y equal the cost of a black ink cartridge. Write a system of equations to represent this situation.

2. What is the cost of each cartridge?

____

3. Ms. Williams is planning to buy T-shirts for the cheerleading camp that she is running. Both companies' total costs would be the same after buying how many T-shirts? Use a graph to find your solution.

	Art Creation Fee	Cost per T-shirt
Company A	$70	$10
Company B	$50	$12

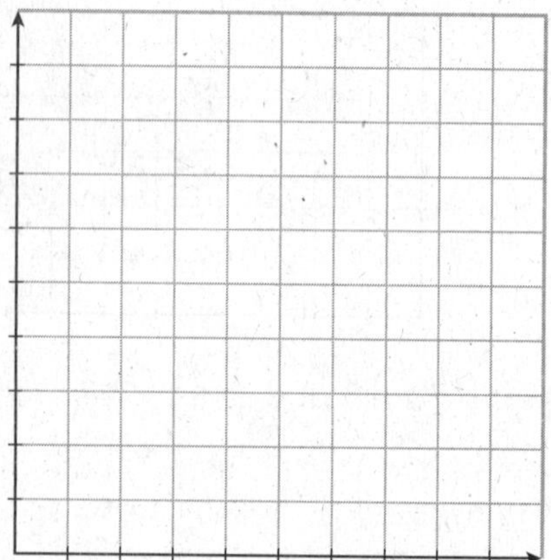

Choose the best answer.

4. Two floats begin a parade at different times, but travel at the same speeds. Which is a true statement about the lines that represent the distance traveled by each float at a given time?

   A The lines intersect.

   B The lines are parallel.

   C The lines are the same.

   D The lines have a negative slope.

5. A piano teacher charges $20 for each half hour lesson, plus an initial fee of $50. Another teacher charges $40 per hour, plus a fee of $50. Which is a true statement about the lines that represent the total cost by each piano teacher?

   F The lines intersect.

   G The lines are parallel.

   H The lines are the same.

   J The lines have a negative slope.

6. Serina is trying to decide between two similar packages for starting her own Web site. Which is a true statement?

   A Both packages cost $235.50 for 5 months.

   B Both packages cost $295 for 10 months.

   C Both packages cost $355 for 15 months.

   D The packages will never have the same cost.

	Design and Setup	Monthly Fee to Host
Package A	$150.00	$14.50
Package B	$175.00	$12.00

Holt Geometry

Problem Solving
Classifying Triangles

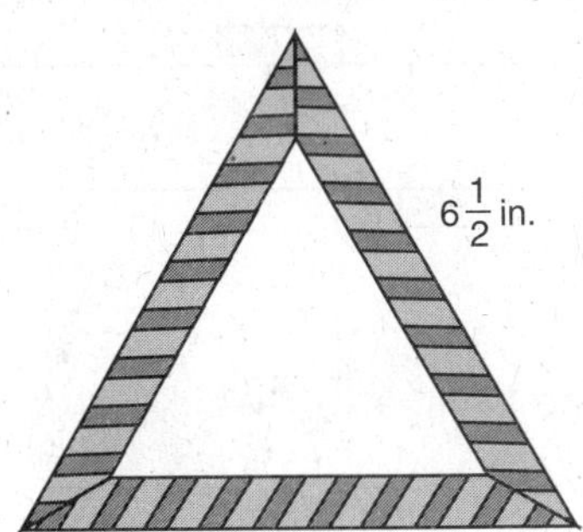

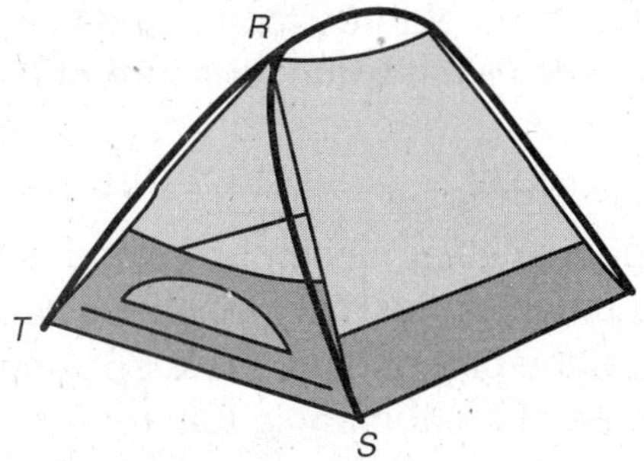

1. Aisha makes triangular picture frames by gluing three pieces of wood together in the shape of an equilateral triangle and covering the wood with ribbon. Each side of a frame is $6\frac{1}{2}$ inches long. How many frames can she cover with 2 yards of ribbon?

2. A tent's entrance is in the shape of an isosceles triangle in which $\overline{RT} \cong \overline{RS}$. The length of $\overline{TS}$ is 1.2 times the length of a side. The perimeter of the entrance is 14 feet. Find each side length.

Use the figure and the following information for Exercises 3 and 4.

The distance "as the crow flies" between Santa Fe and Phoenix is 609 kilometers. This is 245 kilometers less than twice the distance between Santa Fe and El Paso. Phoenix is 48 kilometers closer to El Paso than it is to Santa Fe.

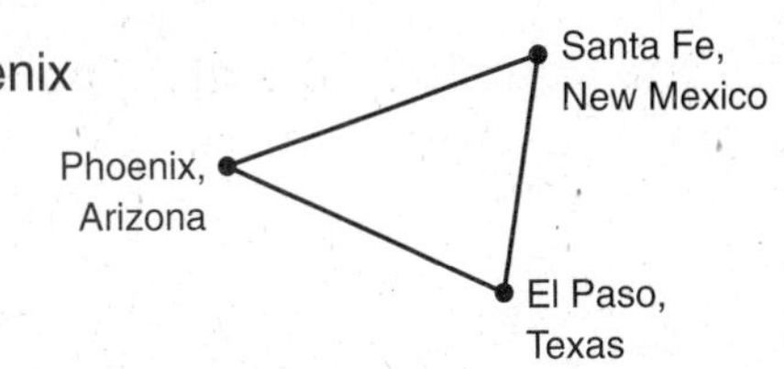

3. What is the distance between each pair of cities?

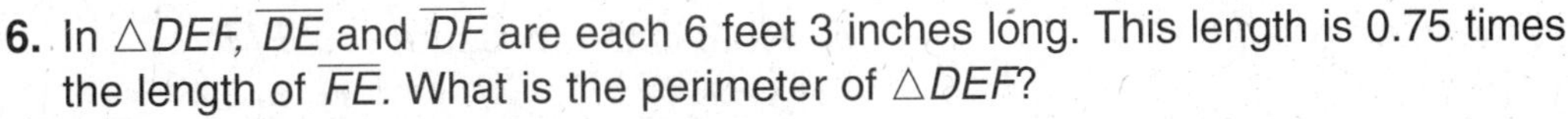

4. Classify the triangle that connects the cities by its side lengths. _______________

Choose the best answer.

A *gable,* as shown in the diagram, is the triangular portion of a wall between a sloping roof.

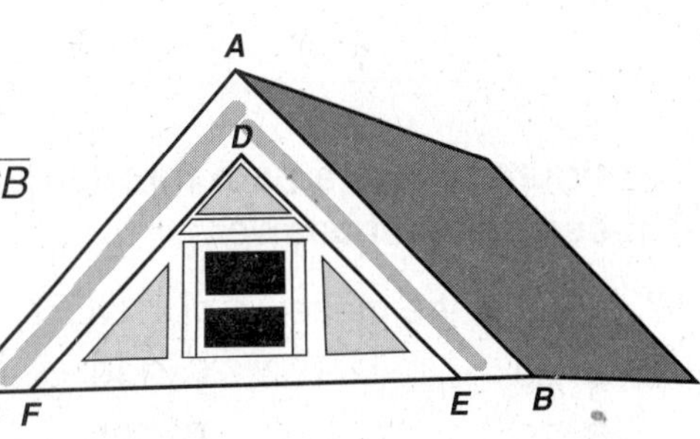

5. Triangle *ABC* is an isosceles triangle. The length of $\overline{CB}$ is 12 feet 4 inches and the congruent sides are each $\frac{3}{4}$ this length. What is the perimeter of $\triangle ABC$?

  A 31 ft 4 in.   **C** 21 ft 7 in.

  B 30 ft 10 in.   **D** 18 ft 6 in.

6. In $\triangle DEF$, $\overline{DE}$ and $\overline{DF}$ are each 6 feet 3 inches long. This length is 0.75 times the length of $\overline{FE}$. What is the perimeter of $\triangle DEF$?

  F 12 ft 4 in.   **H** 17 ft 2 in.

  G 14 ft 7 in.   **J** 20 ft 10 in.

Holt Geometry

Problem Solving
Angle Relationships in Triangles

1. The locations of three food stands on a fair's midway are shown. What is the measure of the angle labeled $x°$?

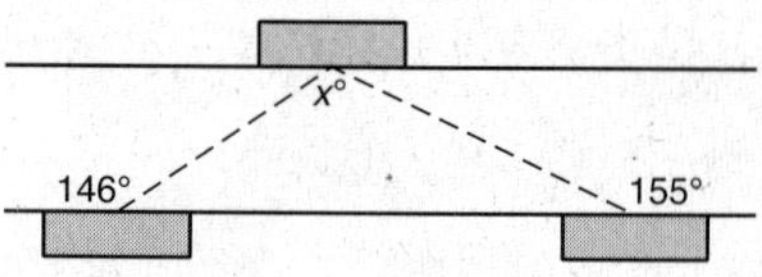

____

2. A large triangular piece of plywood is to be painted to look like a mountain for the spring musical. The angles at the base of the plywood measure 76° and 45°. What is the measure of the top angle that represents the mountain peak?

____

Use the figure of the banner for Exercises 3 and 4.

3. What is the value of n?

____

4. What is the measure of each angle in the banner?

____

Use the figure of the athlete pole vaulting for Exercises 5 and 6.

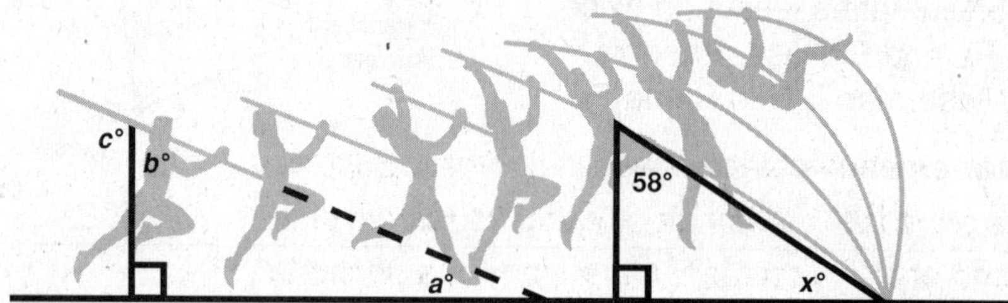

5. What is $x°$, the measure of the angle that the pole makes when it first touches the ground?

____

6. At takeoff, $a° = 23°$. What is $c°$, the measure of the angle the pole makes with the athlete's body?

____

The figure shows a path through a garden. Choose the best answer.

7. What is the measure of $\angle QLP$?

  A 20°   C 110°

  B 70°   D 125°

8. What is the measure of $\angle LPM$?

  F 85°   H 95°

  G 90°   J 125°

9. What is the measure of $\angle PMN$?

  A 98°   C 60°

  B 68°   D 55°

Holt Geometry

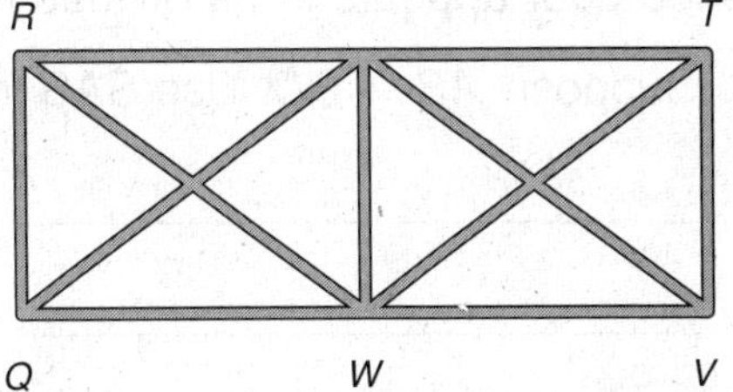

LESSON 4-3
Problem Solving
Congruent Triangles

Use the diagram of the fence for Exercises 1 and 2.

$\triangle RQW \cong \triangle TVW$

1. If $m\angle RWQ = 36°$ and $m\angle TWV = (2x + 5)°$, what is the value of x?

   ____

2. If $RW = (3y - 1)$ feet and $TW = (y + 5)$ feet, what is the length of $\overline{RW}$?

   ____

Use the diagram of a section of the Bank of China Tower for Exercises 3 and 4.

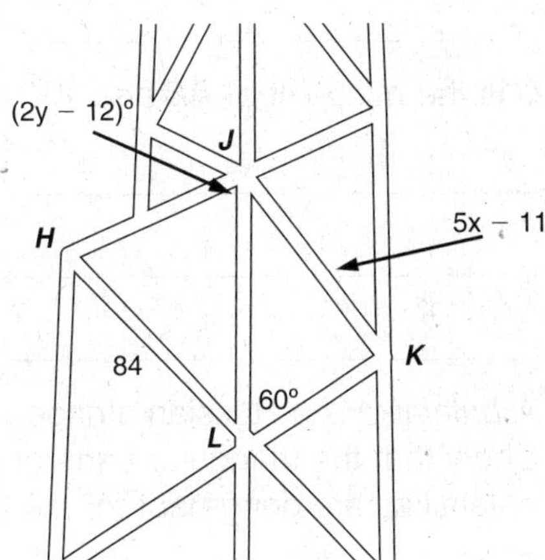

$\triangle JKL \cong \triangle LHJ$

3. What is the value of x?

   ____

4. Find $m\angle JHL$.

   ____

Choose the best answer.

5. Chairs with triangular seats were popular in the Middle Ages. Suppose a chair has a seat that is an isosceles triangle and the congruent sides measure $1\frac{1}{2}$ feet. A second chair has a triangular seat with a perimeter of $5\frac{1}{10}$ feet, and it is congruent to the first seat. What is a side length of the second seat?

   A $1\frac{4}{5}$ ft   **C** 3 ft

   B $2\frac{1}{10}$ ft   **D** $3\frac{3}{5}$ ft

Use the diagram for Exercises 6 and 7.

6. C is the midpoint of $\overline{EB}$ and $\overline{AD}$. What additional information would allow you to prove $\triangle ABC \cong \triangle DEC$ by the definition of congruent triangles?

   F $\overline{EB} \cong \overline{AD}$ **H** $\angle ECD \cong \angle ACB$

   G $\overline{DE} \cong \overline{AB}$ **J** $\angle A \cong \angle D, \angle B \cong \angle E$

7. If $\triangle ABC \cong \triangle DEC$, $ED = 4y + 2$, and $AB = 6y - 4$, what is the length of $\overline{AB}$?

   A 3   **C** 14

   B 12   **D** 18

Holt Geometry

Problem Solving
Triangle Congruence: SSS and SAS

Use the diagram for Exercises 1 and 2.

A shed door appears to be divided into congruent right triangles.

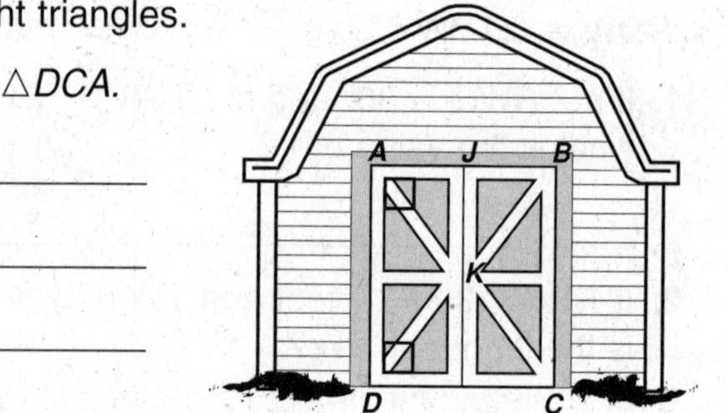

1. Suppose $\overline{AB} \cong \overline{CD}$. Use SAS to show $\triangle ABD \cong \triangle DCA$.

____

____

____

2. J is the midpoint of AB and $\overline{AK} \cong \overline{BK}$. Use SSS to explain why $\triangle AKJ \cong \triangle BKJ$.

____

____

____

3. A *balalaika* is a Russian stringed instrument. Show that the triangular parts of the two balalaikas are congruent for $x = 6$.

____

____

____

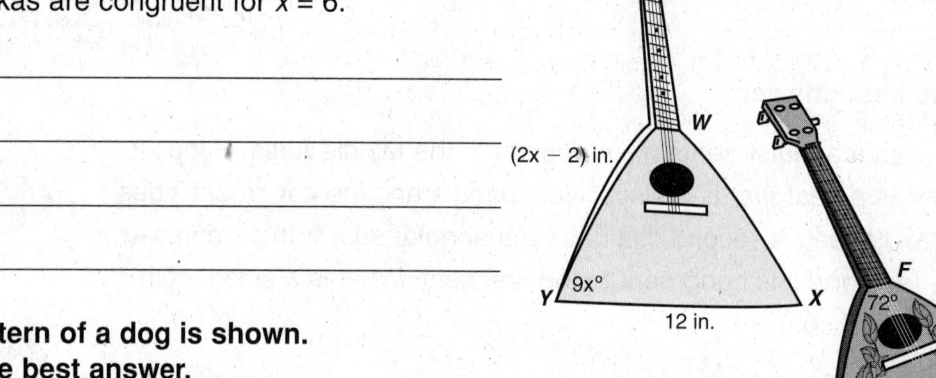

A quilt pattern of a dog is shown. Choose the best answer.

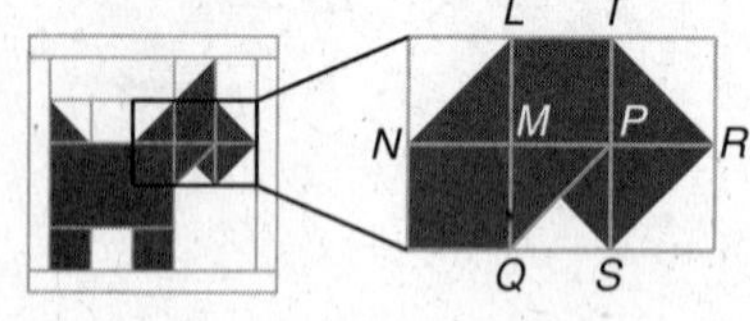

4. $ML = MP = MN = MQ = 1$ inch. Which statement is correct?

   A $\triangle LMN \cong \triangle QMP$ by SAS.

   B $\triangle LMN \cong \triangle QMP$ by SSS.

   C $\triangle LMN \cong \triangle MQP$ by SAS.

   D $\triangle LMN \cong \triangle MQP$ by SSS.

5. P is the midpoint of $\overline{TS}$ and $TR = SR = 1.4$ inches. What can you conclude about $\triangle TRP$ and $\triangle SRP$?

   F $\triangle TRP \cong \triangle SRP$ by SAS.

   G $\triangle TRP \cong \triangle SRP$ by SSS.

   H $\triangle TRP \cong \triangle SPR$ by SAS.

   J $\triangle TRP \cong \triangle SPR$ by SSS.

Holt Geometry

LESSON 4-5 Problem Solving
Triangle Congruence: ASA, AAS, and HL

Use the following information for Exercises 1 and 2.

Melanie is at hole 6 on a miniature golf course. She walks east 7.5 meters to hole 7. She then faces south, turns 67° west, and walks to hole 8. From hole 8, she faces north, turns 35° west, and walks to hole 6.

1. Draw the section of the golf course described. Label the measures of the angles in the triangle.

2. Is there enough information given to determine the location of holes 6, 7, and 8? Explain.

____

3. A section of the front of an English Tudor home is shown in the diagram. If you know that $\overline{KN} \cong \overline{LN}$ and $\overline{JN} \cong \overline{MN}$, can you use HL to conclude that $\triangle JKN \cong \triangle MLN$? Explain.

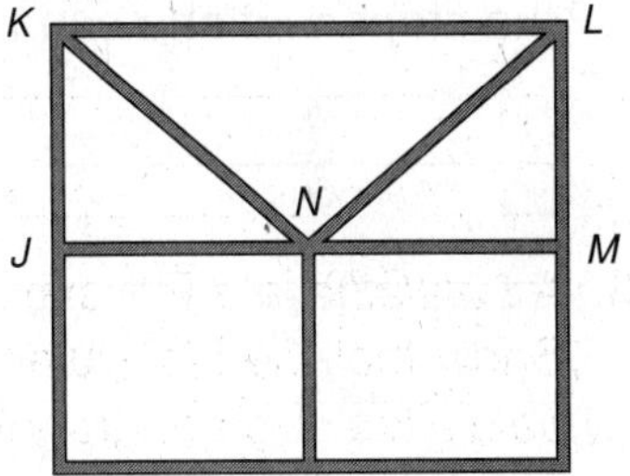

____

____

Use the diagram of a kite for Exercises 4 and 5.

$\overline{AE}$ is the angle bisector of $\angle DAF$ and $\angle DEF$.

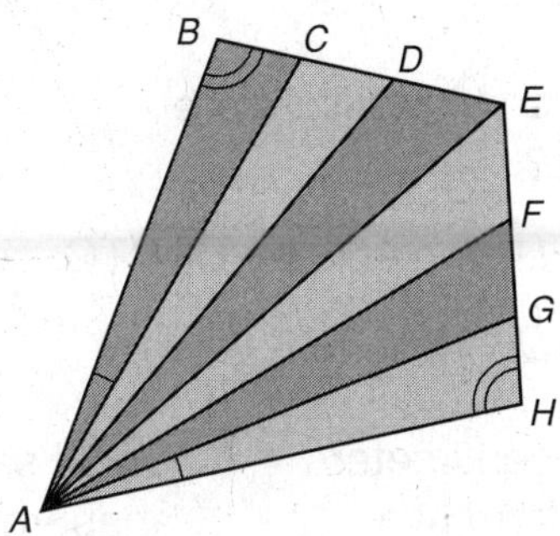

4. What can you conclude about $\triangle DEA$ and $\triangle FEA$?

   A $\triangle DEA \cong \triangle FEA$ by HL.

   B $\triangle DEA \cong \triangle FEA$ by AAA.

   C $\triangle DEA \cong \triangle FEA$ by ASA.

   D $\triangle DEA \cong \triangle FEA$ by SAS.

5. Based on the diagram, what can you conclude about $\triangle BCA$ and $\triangle HGA$?

   F $\triangle BCA \cong \triangle HGA$ by HL.

   G $\triangle BCA \cong \triangle HGA$ by AAS.

   H $\triangle BCA \cong \triangle HGA$ by ASA.

   J It cannot be shown using the given information that $\triangle BCA \cong \triangle HGA$.

Holt Geometry

Problem Solving
Triangle Congruence: CPCTC

1. Two triangular plates are congruent. The area of one of the plates is 60 square inches. What is the area of the other plate? Explain.

____

____

2. An archaeologist draws the triangles to find the distance *XY* across a ravine. What is *XY*? Explain.

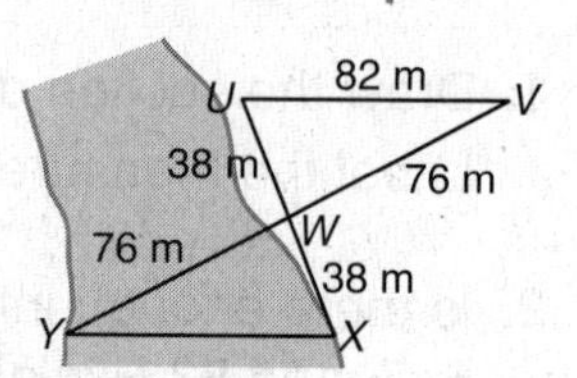

____

____

3. A city planner sets up the triangles to find the distance *RS* across a river. Describe the steps that she can use to find *RS*.

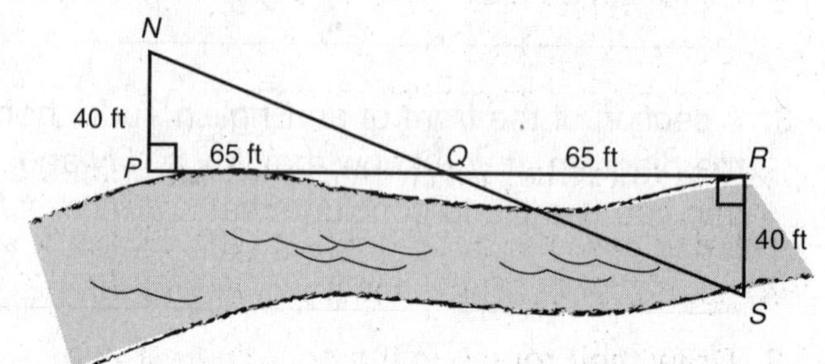

____

____

____

____

____

Choose the best answer.

4. A lighthouse and the range of its shining light are shown. What can you conclude?

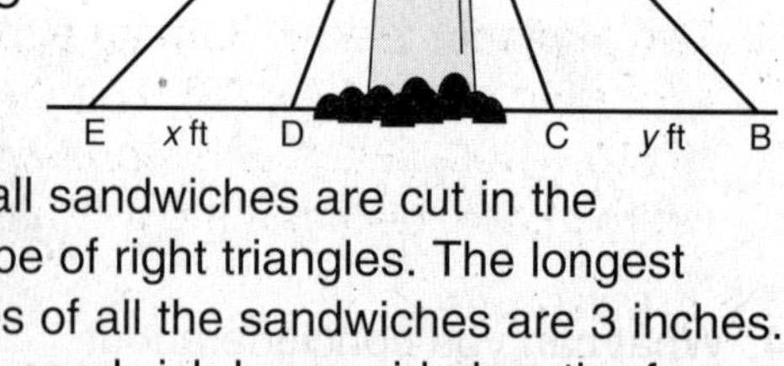

  A $x = y$ by CPCTC   **C** $\angle AED \cong \angle ADE$ by CPCTC

  B $x = 2y$ **D** $\angle AED \cong \angle ACB$

5. A rectangular piece of cloth 15 centimeters long is cut along a diagonal to form two triangles. One of the triangles has a side length of 9 centimeters. Which is a true statement?

  F The second triangle has an angle measure of 15° by CPCTC.

  G The second triangle has a side length of 9 centimeters by CPCTC.

  H You cannot make a conclusion about the side length of the second triangle.

  J The triangles are not congruent.

6. Small sandwiches are cut in the shape of right triangles. The longest sides of all the sandwiches are 3 inches. One sandwich has a side length of 2 inches. Which is a true statement?

  A All the sandwiches have a side length of 2 inches by CPCTC.

  B All the sandwiches are isosceles triangles with side lengths of 2 inches.

  C None of the other sandwiches have side lengths of 2 inches.

  D You cannot make a conclusion using CPCTC.

Holt Geometry

Problem Solving

Introduction to Coordinate Proof

Round to the nearest tenth for Exercises 1 and 2.

1. A fountain is at the center of a square courtyard. If one grid unit represents one yard, what is the distance from the fountain at $(0, 0)$ to each corner of the courtyard?

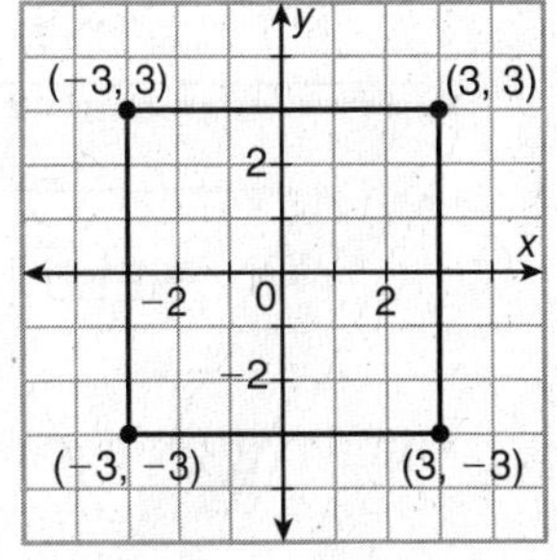

____

2. Noah started at his home at $A(0, 0)$, walked with his dog to the park at $B(4, 2)$, walked to his friend's house at $C(8, 0)$, then walked home. If one grid unit represents 20 meters, what is the distance that Noah and his dog walked?

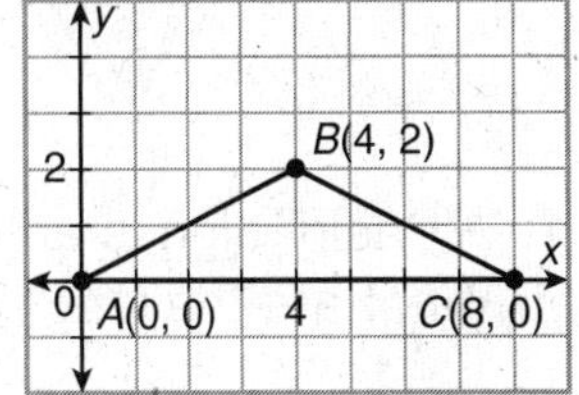

____

Use the following information for Exercises 3 and 4.

Rachel started her cycling trip at $G(0, 7)$. Malik started his trip at $J(0, 0)$. Their paths crossed at $H(4, 2)$.

3. Draw their routes in the coordinate plane.

4. If one grid unit represents $\frac{1}{2}$ mile, who had ridden farther when their paths crossed? Explain.

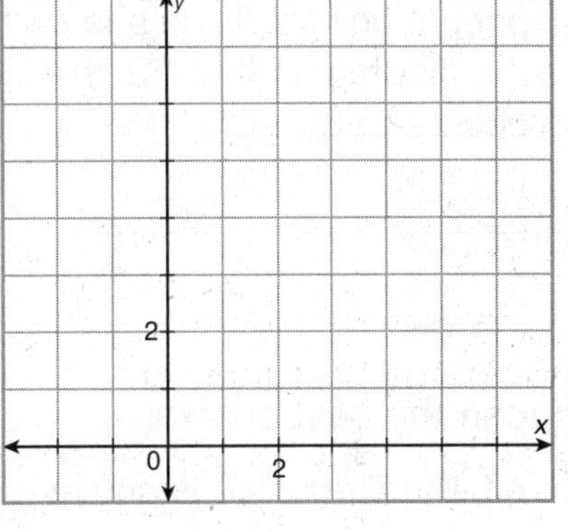

____

____

Choose the best answer.

5. Two airplanes depart from an airport at $A(9, 11)$. The first airplane travels to a location at $N(-250, 80)$, and the second airplane travels to a location at $P(105, -400)$. Each unit represents 1 mile. What is the distance, to the nearest mile, between the two airplanes?

   A 335.3 mi

   B 477.9 mi

   C 490.3 mi

   D 597.0 mi

6. A corner garden has vertices at $Q(0, 0)$, $R(0, 2d)$, and $S(2c, 0)$. A brick walkway runs from point Q to the midpoint M of $\overline{RS}$. What is QM?

   F (c, d)

   G $c^2 + d^2$

   H $\sqrt{c + d}$

   J $\sqrt{c^2 + d^2}$

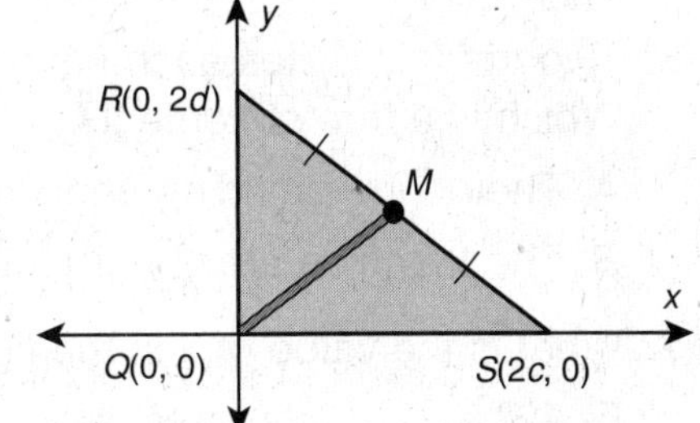

Holt Geometry

LESSON 4-8
Problem Solving
Isoceles and Equilateral Triangles

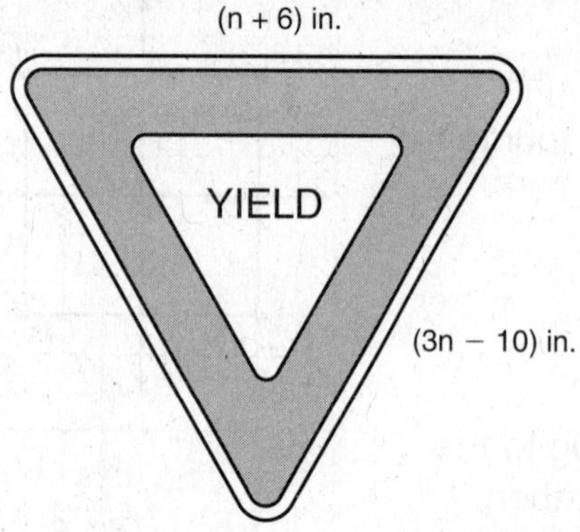

1. A "Yield" sign is an equiangular triangle. What are the lengths of the sides?

2. The measure of $\angle C$ is 70°. What is the measure of $\angle B$?

____

3. Samantha is swimming along $\overrightarrow{HF}$. When she is at point H, she sees a necklace straight ahead of her but on the bottom of the pool at point J. Then she swims 11 more feet to point G. Use the diagram to find GJ, the distance Samantha is from the necklace. Explain.

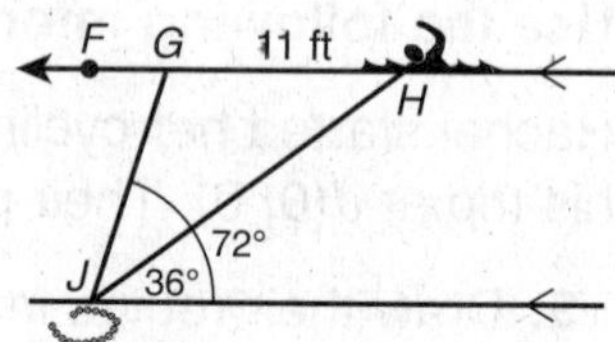

____

____

Choose the best answer.

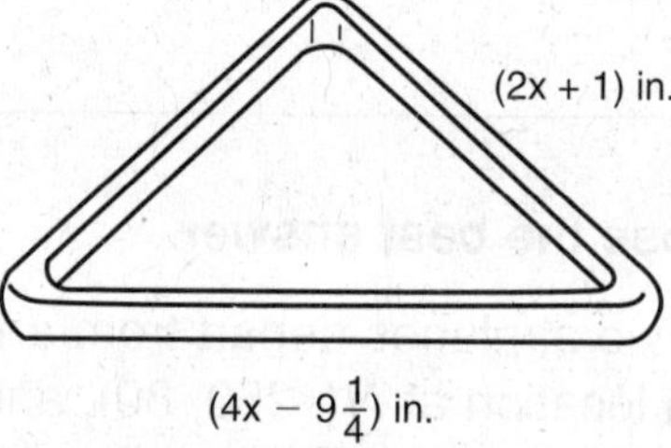

4. A billiards triangle is equiangular. What is the perimeter?

  A $5\frac{1}{8}$ in.   **C** $11\frac{1}{4}$ in.

  B $10\frac{1}{4}$ in.   **D** $33\frac{3}{4}$ in.

5. A triangular shaped trellis has angles R, S, and T that measure 73°, 73°, and 34°, respectively. If $ST = 4y + 6$ and $TR = 7y − 21$, what is the value of y?

  F 5   **H** 11

  G 9   **J** 15

6. Two triangular tiles each have two sides measuring 4 inches. Which is a true statement?

  A Their corresponding angles are congruent.   **C** The triangles may be congruent.

  B The triangles are congruent.   **D** The triangles cannot be congruent.

7. What is the value of x in the figure?

  F 42°   **H** 96°

  G 90°   **J** 106°

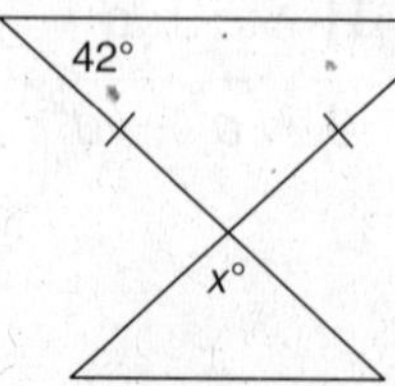

Holt Geometry

Problem Solving
Perpendicular and Angle Bisectors

Use the diagram for Exercises 1 and 2.

Fire stations are located at *A* and *B*. $\overleftrightarrow{XY}$, which contains Havens Road, represents the perpendicular bisector of $\overline{AB}$.

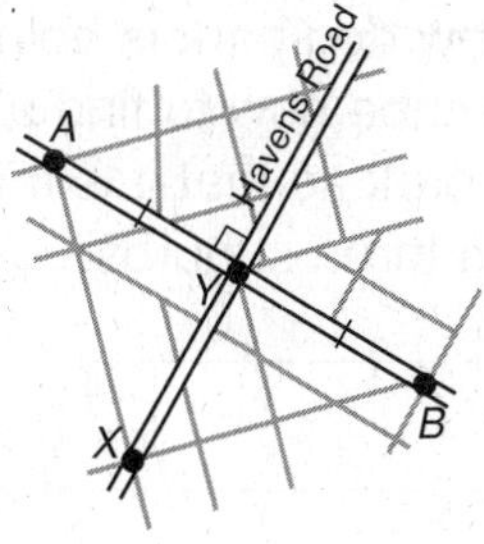

1. A fire is reported at point *X*. Which fire station is closer to the fire? Explain.

____

____

____

2. The city wants to build a third fire station so that it is the same distance from the stations at *A* and *B*. How can the city be sure that this is the case?

____

____

____

3. Wire is used to hang the picture on a nail at point *S*. How can the two lengths of wire, *SR* and *ST*, be used so that the picture is straight and centered under the nail?

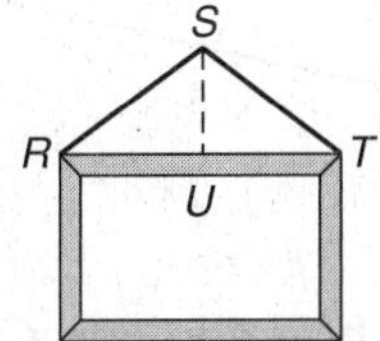

4. A piece of wood for a birdhouse is shown. Point *H* is the center of a ventilation hole that is to be drilled 2 inches from $\overline{FE}$ and $\overrightarrow{FG}$. If you drew $\overrightarrow{FH}$, what would be $m\angle EFH$? Explain.

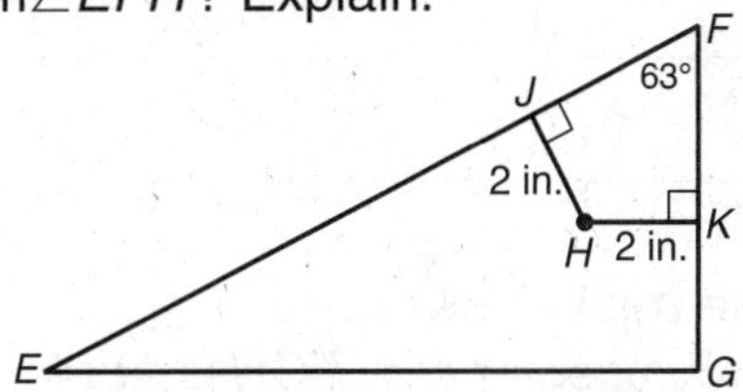

____

____

____

Choose the best answer.

The design at the right was made by wrapping string around nails.

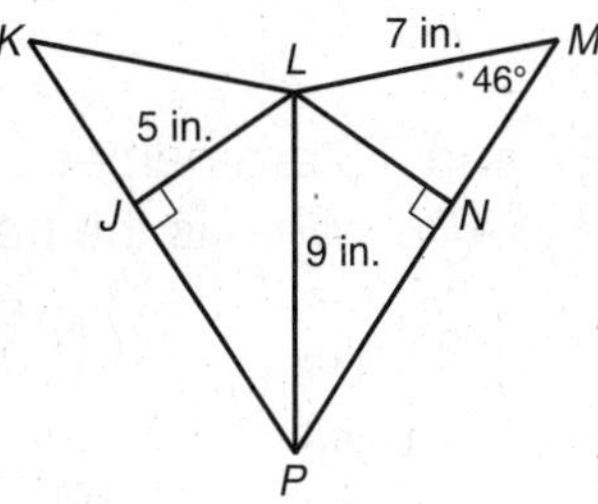

5. $\overline{PL}$ is the angle bisector of $\angle KPM$. Which can you conclude from this statement?

   A $LN = 5$ in.   **C** $m\angle K = 46°$

   B $LK = 7$ in.   **D** $m\angle JLK = 44°$

6. $\overline{LJ}$ is the perpendicular bisector of $\overline{KP}$. Which can you conclude?

   F $m\angle K = 46°$ **H** $KL = 9$ in.

   G $m\angle K = 44°$ **J** $KL = 7$ in.

Holt Geometry

LESSON 5-2 Problem Solving
Bisectors of Triangles

1. A new dog park is being planned. Describe how to find a location for the park so that it is the same distance from three suburbs.

____

____

____

____

2. A fountain is in a triangular sitting area of a mall, △ABC. A diagram shows that the fountain is at the point where the angle bisectors of △ABC are concurrent. If the distance from the fountain to one wall is 15 feet, what is the distance from the fountain to another wall? Explain.

____

____

____

3. A water tower is to be built so that it is the same distance from the cities at X, Y, and Z. Draw a sketch on △XYZ to show the location W where the water tower should be built. Justify your sketch.

____

____

____

____

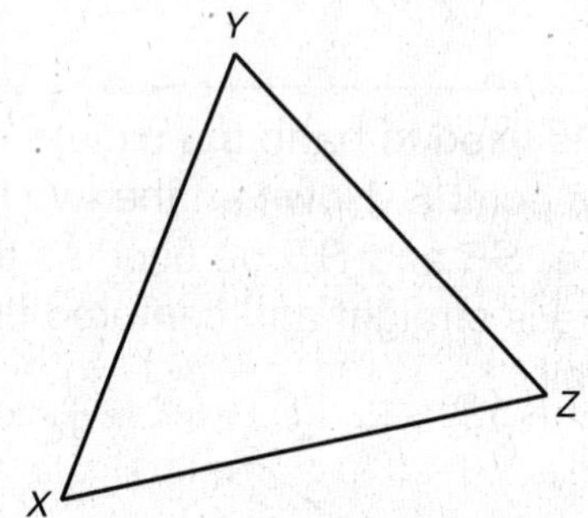

Choose the best answer.

4. The circumcenter of △FGH is at (4, −5). If G is at (0, 0), which of the following are possible coordinates of F and H?

   A F(0, −8), H(10, 0)

   B F(0, 8), H(−10, 0)

   C F(0, −10), H(8, 0)

   D F(0, 10), H(−8, 0)

5. A triangle has vertices Q(−9, 10), R(0, 1), and S(8, 4). Which is a correct statement about the incenter and circumcenter of △QRS?

   F Both points are on △QRS.

   G Both points are inside △QRS.

   H Both points are outside △QRS.

   J One point is inside △QRS, and one point is outside △QRS.

6. $\overline{RT}$ and $\overline{TS}$ are perpendicular bisectors of △ABC. What is the perimeter of △ATC?

   A 17.2 units

   B 19.4 units

   C 20.9 units

   D 22.4 units

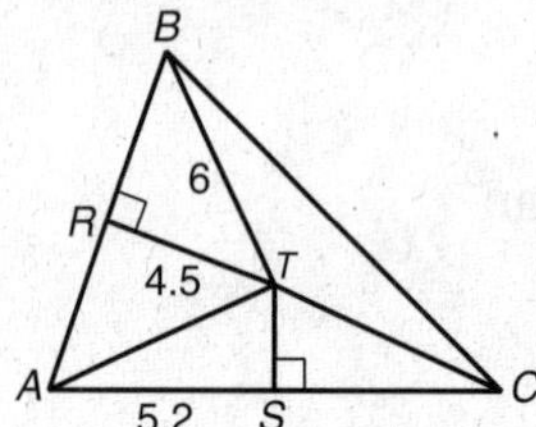

7. If m∠KPN = 44°, find m∠JLP.

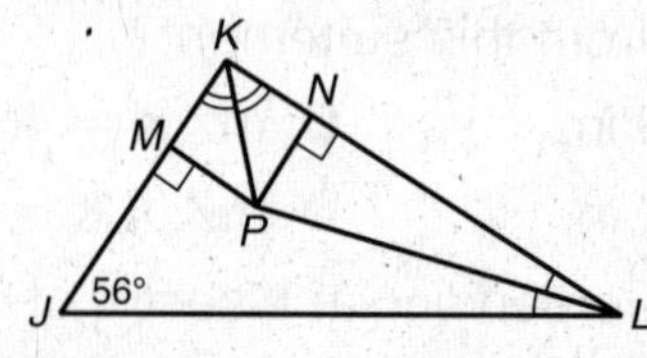

   F 16°

   G 18°

   H 23°

   J 32°

Holt Geometry

Problem Solving
Medians and Altitudes of Triangles

1. The diagram shows the coordinates of the vertices of a triangular patio umbrella. The umbrella will rest on a pole that will support it. Where should the pole be attached so that the umbrella is balanced?

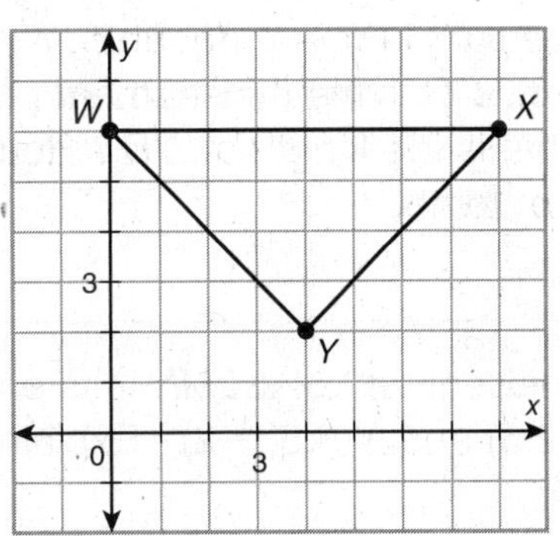

____

2. In a plan for a triangular wind chime, the coordinates of the vertices are $J(10, 2)$, $K(7, 6)$, and $L(12, 10)$. At what coordinates should the manufacturer attach the chain from which it will hang in order for the chime to be balanced?

3. Triangle PQR has vertices at $P(-3, 5)$, $Q(-1, 7)$, and $R(3, 1)$. Find the coordinates of the orthocenter and the centroid.

Choose the best answer.

4. A triangle has coordinates at $A(0, 6)$, $B(8, 6)$, and $C(5, 0)$. $\overline{CD}$ is a median of the triangle, and $\overline{CE}$ is an altitude of the triangle. Which is a true statement?

   A The coordinates of D and E are the same.

   B The distance between D and E is 1 unit.

   C The distance between D and E is 2 units.

   D D is on the triangle, and E is outside the triangle.

5. Lines j and k contain medians of $\triangle DEF$. Find y and z.

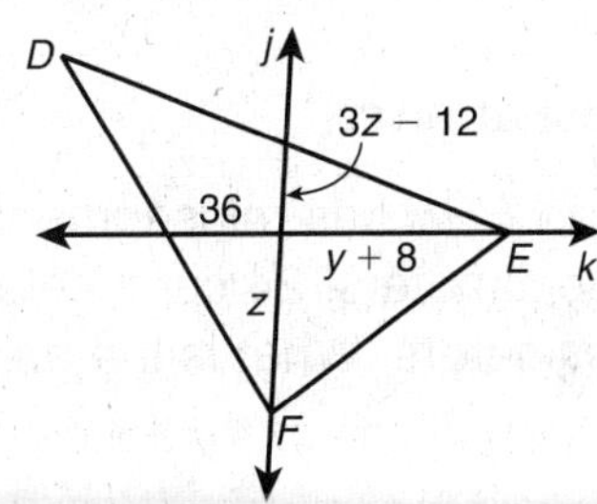

   F $y = 16$; $z = 4$ H $y = 64$; $z = 4.8$

   G $y = 32$; $z = 4$ J $y = 108$; $z = 8$

6. An inflatable triangular raft is towed behind a boat. The raft is an equilateral triangle. To maintain balance, the seat is at the centroid B of the triangle. What is AB, the distance from the seat to the tow rope? Round to the nearest tenth.

   A 18.7 in.

   B 37.4 in.

   C 43.1 in.

   D 56.0 in.

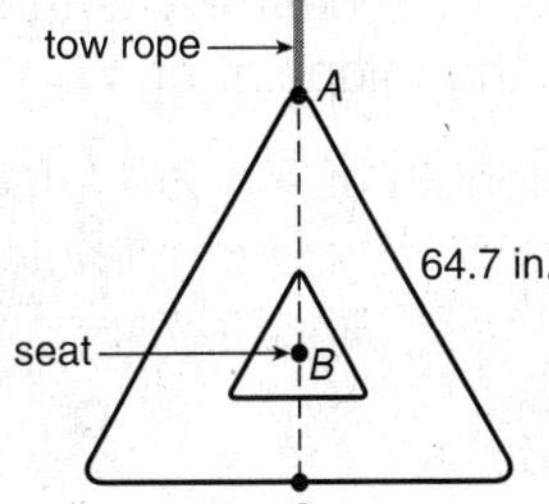

Holt Geometry

Problem Solving
LESSON 5-4

The Triangle Midsegment Theorem

1. The vertices of $\triangle JKL$ are $J(-9, 2)$, $K(10, 1)$, and $L(5, 6)$. $\overline{CD}$ is the midsegment parallel to $\overline{JK}$. What is the length of $\overline{CD}$? Round to the nearest tenth.

2. In $\triangle QRS$, $QR = 2x + 5$, $RS = 3x - 1$, and $SQ = 5x$. What is the perimeter of the midsegment triangle of $\triangle QRS$?

3. Is XY a midsegment of $\triangle LMN$ if its endpoints are $X(8, 2.5)$ and $Y(6.5, -2)$? Explain.

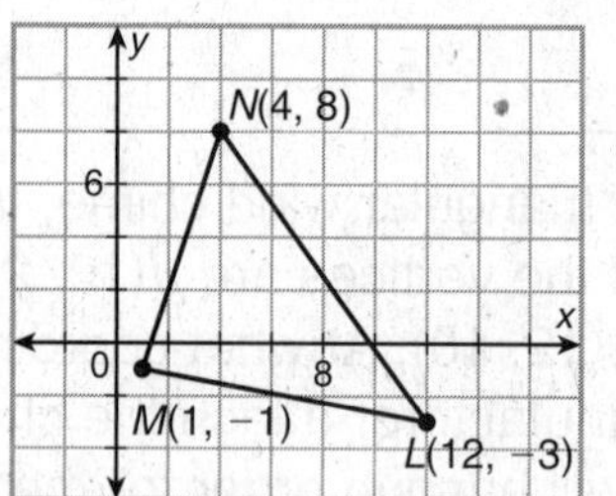

4. The diagram at right shows horseback riding trails. Point B is the halfway point along path $\overline{AC}$. Point D is the halfway point along path $\overline{CE}$. The paths along $\overline{BD}$ and $\overline{AE}$ are parallel. If riders travel from A to B to D to E, and then back to A, how far do they travel?

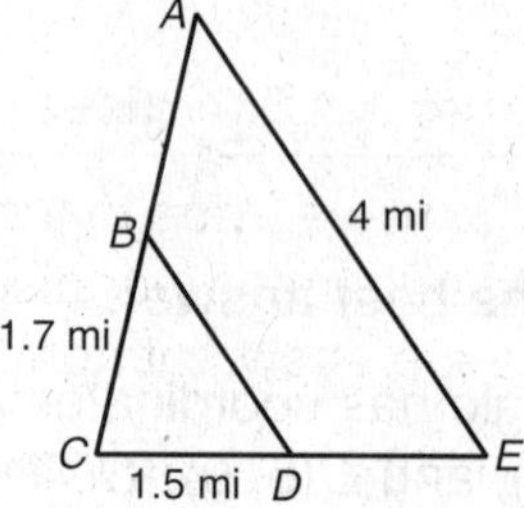

Choose the best answer.

5. Right triangle FGH has midsegments of length 10 centimeters, 24 centimeters, and 26 centimeters. What is the area of $\triangle FGH$?

  A 60 cm^2 **C** 240 cm^2

  B 120 cm^2 **D** 480 cm^2

6. In triangle HJK, $m\angle H = 110°$, $m\angle J = 30°$, and $m\angle K = 40°$. If R is the midpoint of $\overline{JK}$, and S is the midpoint of $\overline{HK}$, what is $m\angle JRS$?

  F 150°   **H** 110°

  G 140°   **J** 30°

Use the diagram for Exercises 7 and 8.

On the balance beam, V is the midpoint of $\overline{AB}$, and W is the midpoint of $\overline{YB}$.

7. The length of $\overline{VW}$ is $1\frac{7}{8}$ feet. What is AY?

  A $\frac{7}{8}$ ft   **C** $3\frac{3}{4}$ ft

  B $\frac{15}{16}$ ft   **D** $7\frac{1}{2}$ ft

8. The measure of $\angle AYW$ is 50°. What is the measure of $\angle VWB$?

  F 45°   **H** 90°

  G 50°   **J** 130°

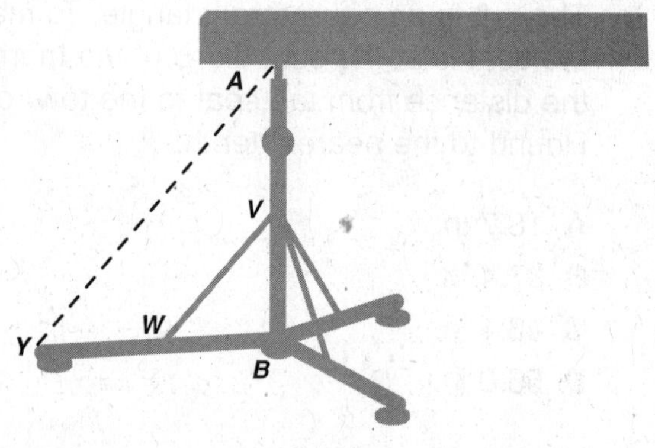

Holt Geometry

Problem Solving

LESSON 5-5

Indirect Proof and Inequalities in One Triangle

1. A charter plane travels from Barrow, Alaska, to Fairbanks. From Fairbanks, it flies to Nome, and then back to its starting point in Barrow. Which of the three legs of the trip is the longest?

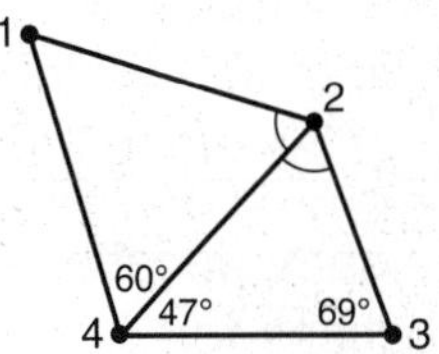

____

2. Three cell phone towers are shown at the right. The measure of $\angle M$ is 10° less than the measure of $\angle K$. The measure of $\angle L$ is 1° greater than the measure of $\angle K$. Which two towers are closest together?

____

Use the figure for Exercises 3 and 4.

In disc golf, a player tries to throw a disc into a metal basket target. Four disc golf targets on a course are shown at right.

3. Which two targets are closest together?

4. Which two targets are farthest apart?

___________________________________   ___________________________________

Choose the best answer.

5. The distance from Jacksonville to Tampa is 171 miles. The distance from Tampa to Miami is 206 miles. Use the Triangle Inequality Theorem to find the range for the distance from Jacksonville to Miami.

   A $0\text{ mi} < d < 35\text{ mi}$

   B $0\text{ mi} < d < 377\text{ mi}$

   C $35\text{ mi} < d < 377\text{ mi}$

   D $-35\text{ mi} < d < 377\text{ mi}$

6. In Jessica's room, the distance from the door D to the closet C is 4 feet. The distance from the closet to the window W is 6 feet. The distance from the window to the door is 8 feet. On a floor plan of her room, $\triangle CDW$ is drawn. Order the angles from least to greatest measure.

   F $\angle C, \angle D, \angle W$ **H** $\angle W, \angle C, \angle D$

   G $\angle D, \angle C, \angle W$ **J** $\angle W, \angle D, \angle C$

7. Walking paths at a park are shown. Which route represents the greatest distance?

   A A to B to D **C** C to B to D

   B A to D to B **D** C to D to B

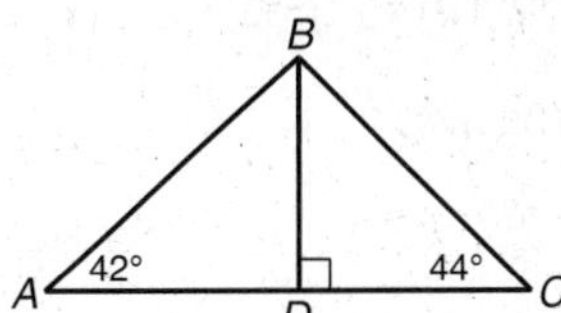

Holt Geometry

Problem Solving
Inequalities in Two Triangles

1. The angle that a person makes as he or she is sitting changes with the task. The diagram shows the position of a student at his desk. In which position is the angle measure $a°$ at which he is sitting the greatest? The least? Explain.

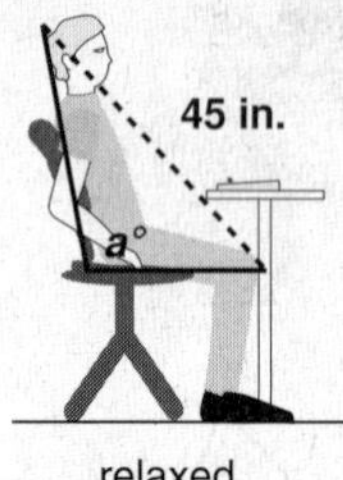

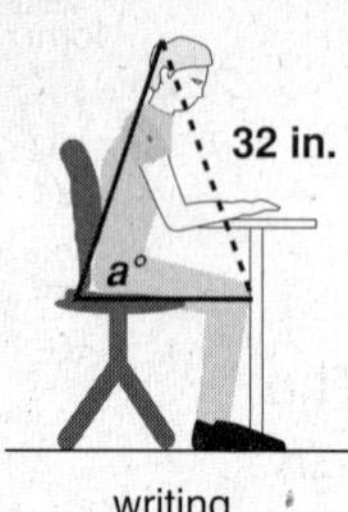

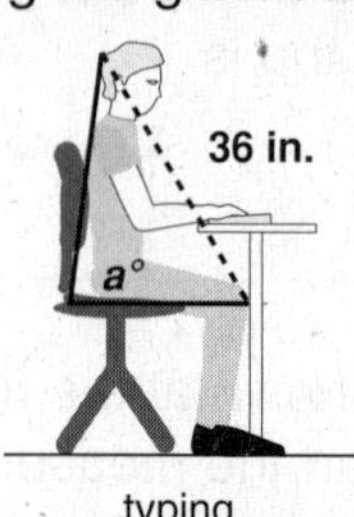

____

____

____

____

2. Two cyclists start from the same location and travel in opposite directions for 2 miles each. Then the first cyclist turns right 90° and continues for another mile. At the same time, the second cyclist turns 45° left and continues for another mile. At this point, which cyclist is closer to the original starting point?

____

3. A compass is used to draw a circle. Then the compass is opened wider and another circle is drawn. Explain how this illustrates the Hinge Theorem.

____

____

____

____

____

Choose the best answer.

4. Two sides of each triangle in the circle are formed from the radii of the circle. Compare EF and FG.

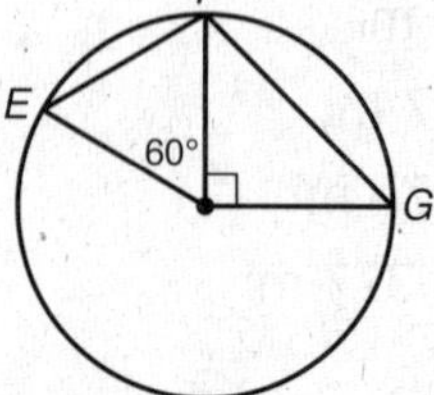

  A $EF = FG$

  B $EF < FG$

  C $EF > FG$

  D Not enough information is given.

5. Compare $m\angle Y$ and $m\angle M$.

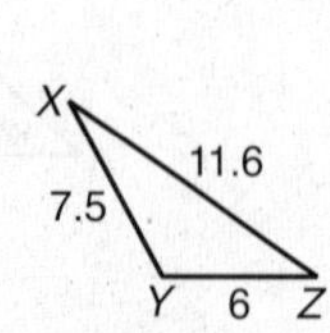

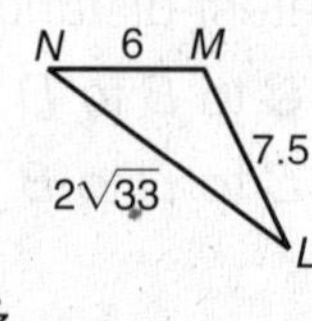

  F $m\angle Y = m\angle M$

  G $m\angle Y > m\angle M$

  H $m\angle Y < m\angle M$

  J Not enough information is given.

Holt Geometry

Problem Solving
The Pythagorean Theorem

1. It is recommended that for a height of 20 inches, a wheelchair ramp be 19 feet long. What is the value of x to the nearest tenth?

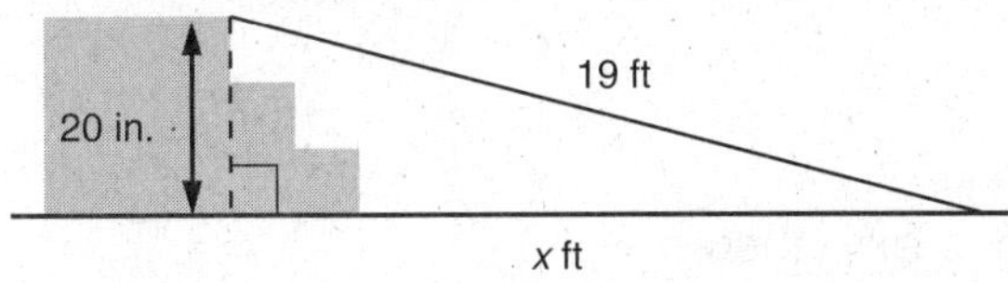

2. Find x, the length of the weight-lifting incline bench. Round to the nearest tenth.

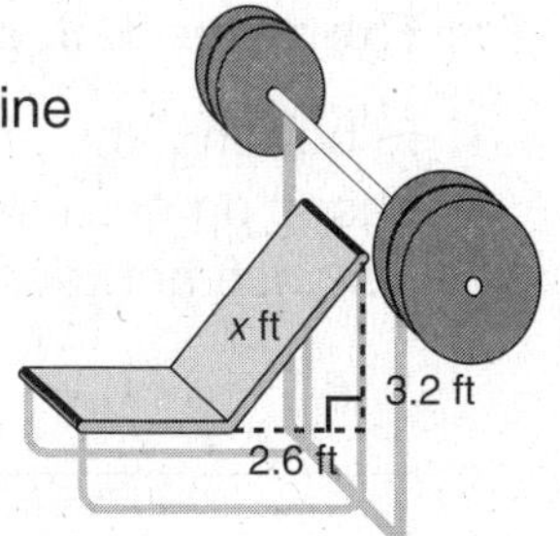

3. A ladder 15 feet from the base of a building reaches a window that is 35 feet high. What is the length of the ladder to the nearest foot?

4. In a wide-screen television, the ratio of width to height is 16 : 9. What are the width and height of a television that has a diagonal measure of 42 inches? Round to the nearest tenth.

Choose the best answer.

5. The distance from Austin to San Antonio is about 74 miles, and the distance from San Antonio to Victoria is about 102 miles. Find the approximate distance from Austin to Victoria.

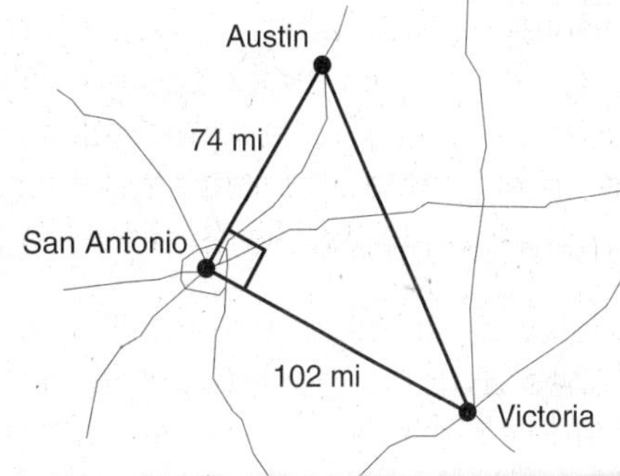

A 28 mi   **C** 126 mi

B 70 mi   **D** 176 mi

6. What is the approximate perimeter of $\triangle DEC$ if rectangle $ABCD$ has a length of 4.6 centimeters?

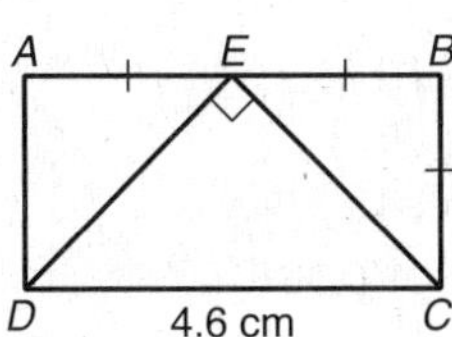

F 5.1 cm

G 6.5 cm

H 9.8 cm

J 11.1 cm

7. The legs of a right triangle measure $3x$ and 15. If the hypotenuse measures $3x + 3$, what is the value of x?

A 12   **C** 36

B 16   **D** 221

8. A cube has edge lengths of 6 inches. What is the approximate length of a diagonal d of the cube?

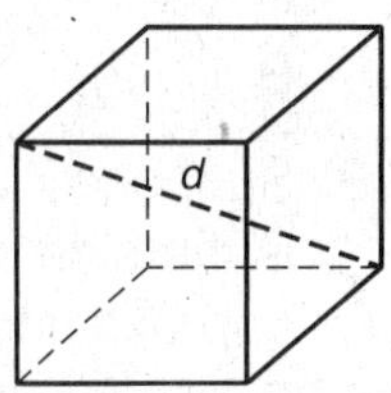

F 6 in.   **H** 10.4 in.

G 8.4 in.   **J** 12 in.

Holt Geometry

LESSON 5-8 Problem Solving
Applying Special Right Triangles

For Exercises 1–6, give your answers in simplest radical form.

1. In bowling, the pins are arranged in a pattern based on equilateral triangles. What is the distance between pins 1 and 5?

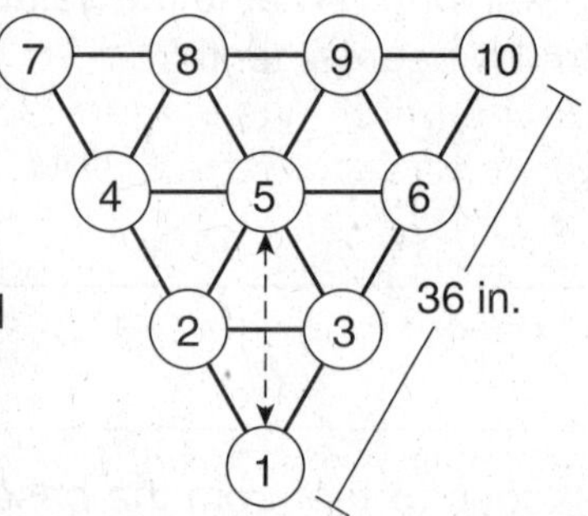

____

2. To secure an outdoor canopy, a 64-inch cord is extended from the top of a vertical pole to the ground. If the cord makes a 60° angle with the ground, how tall is the pole?

____

Find the length of $\overline{AB}$ in each quilt pattern.

3. 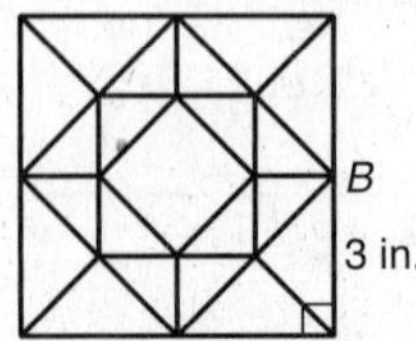

4. 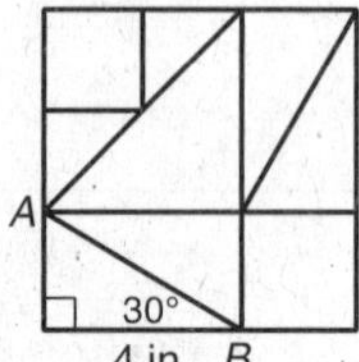

____

____

Choose the best answer.

5. An equilateral triangle has an altitude of 21 inches. What is the side length of the triangle?

6. A shelf is an isosceles right triangle, and the longest side is 38 centimeters. What is the length of each of the other two sides?

____

____

Use the figure for Exercises 7 and 8.

Assume $\triangle JKL$ is in the first quadrant, with $m\angle K = 90°$.

7. Suppose that $\overline{JK}$ is a leg of $\triangle JKL$, a 45°-45°-90° triangle. What are possible coordinates of point L?

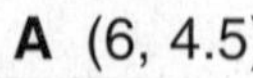

   A (6, 4.5)   C (6, 2)

   B (7, 2)   D (8, 7)

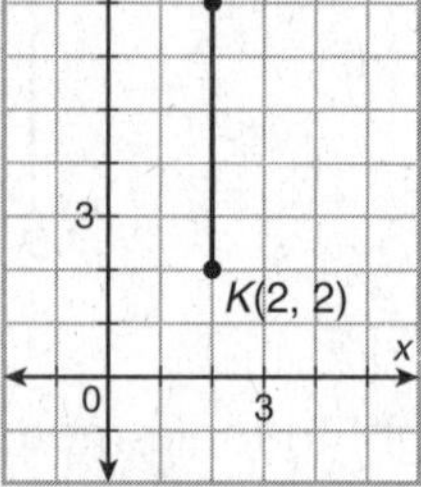

8. Suppose $\triangle JKL$ is a 30°-60°-90° triangle and $\overline{JK}$ is the side opposite the 60° angle. What are the approximate coordinates of point L?

   F (4.9, 2)   H (8.7, 2)

   G (4.5, 2)   J (7.1, 2)

Holt Geometry

Problem Solving
Properties and Attributes of Polygons

1. A campground site is in the shape of a convex quadrilateral. Three sides of the campground form two right angles. The third interior angle measures 10° less than the fourth angle. Find the measure of each interior angle.

____

2. A pentagon has two exterior angles that measure $(3x)°$, two exterior angles that measure $(2x + 22)°$, and an exterior angle that measures $(x + 41)°$. If all of these angles have different vertices, what are the measures of the exterior angles of the pentagon?

____

3. The top view of a hexagonal greenhouse is shown at the right. What is the measure of $\angle PQR$, the acute angle formed by the house and the greenhouse?

____

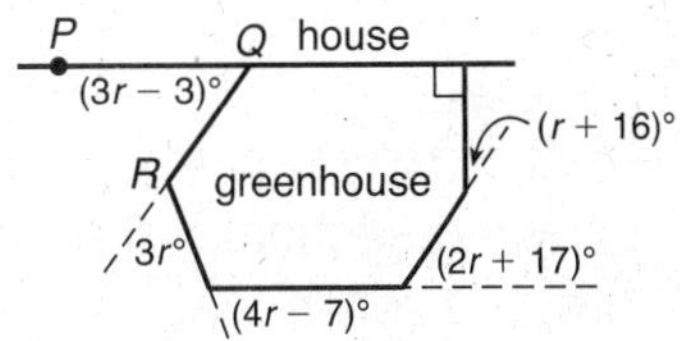

Choose the best answer.

4. A figure is an equiangular 18-gon. What is the measure of each exterior angle of the polygon?

   A 10°

   B 18°

   C 20°

   D 36°

5. Three interior angles of a convex heptagon measure 125°, and two of the interior angles measure 143°. Which are possible measures for the other two interior angles of the heptagon?

   F 48° and 48°   **H** 100° and 116°

   G 39° and 100°   **J** 89° and 150°

6. Find the measure of $\angle RKL$.

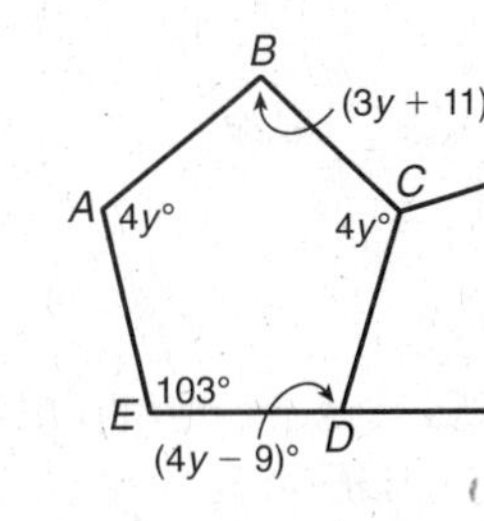

   A 34°   **C** 86°

   B 68°   **D** 148°

7. What is the measure of $\angle GCD$?

   F 123°   **H** 73°

   G 116°   **J** 29°

Holt Geometry

LESSON 6-2

Problem Solving
Properties of Parallelograms

Use the diagram for Exercises 1 and 2.
The wall frames on the staircase wall form parallelograms
ABCD and *EFGH*.

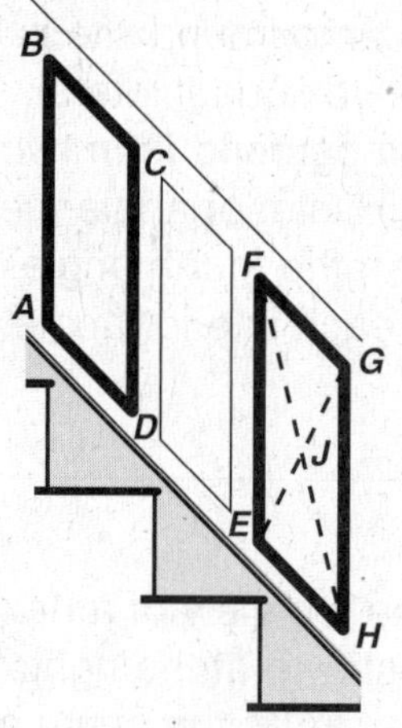

1. In $\square ABCD$, the measure of $\angle A$ is three times the measure
of $\angle B$. What are the measures of $\angle C$ and $\angle D$?

____

2. In $\square EFGH$, $FH = 5x$ inches, $EG = (2x + 4)$ inches,
and $JG = 8$ inches. What is the length of JH?

____

3. The diagram shows a section of the
support structure of a roller coaster.
In $\square JKLM$, $JK = (3z - 0.9)$ feet, and
$LM = (z + 2.7)$ feet. Find JK.

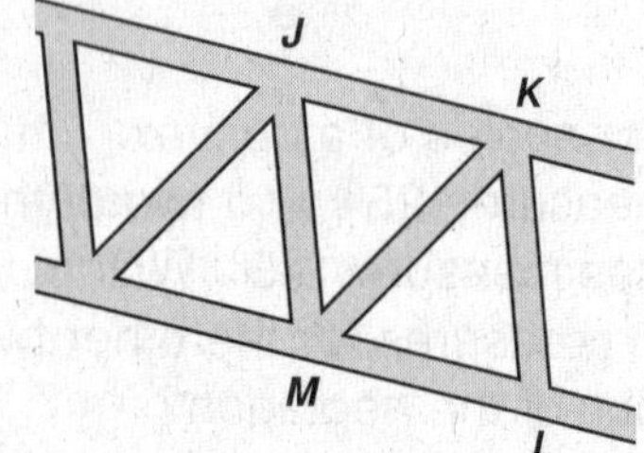

4. In $\square TUVW$, part of a ceramic tile
pattern, $m\angle TUV = (8x + 1)°$ and
$m\angle UVW = (12x + 19)°$. Find $m\angle TUV$.

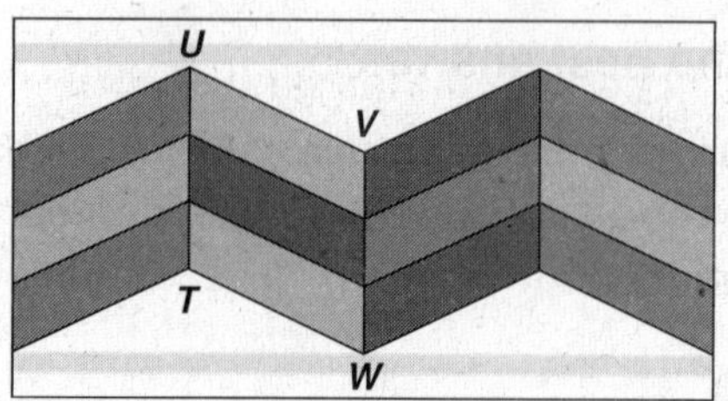

Choose the best answer.

5. What is the measure of $\angle Z$ in
parallelogram *WXYZ*?

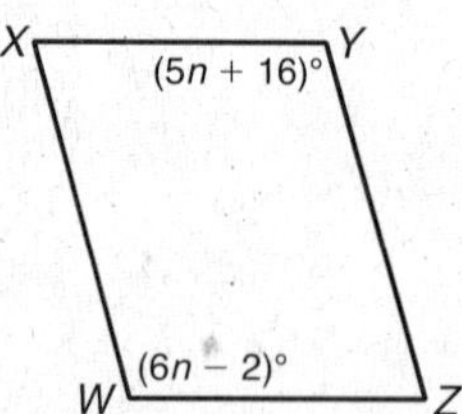

  A 18°

  B 74°

  C 106°

  D 108°

6. The perimeter of $\square CDEF$ is 54 centimeters.
Find the length of $\overline{FC}$ if $\overline{DE}$ is 5 centimeters
longer than $\overline{EF}$.

  F 11 cm

  G 14 cm

  H 16 cm

  J 44 cm

7. In $\square PQRS$, $QT = 7x$, $TS = 2x + 2.5$,
$RT = 2y$, and $TP = y + 3$. Find the
perimeter of $\triangle PTS$.

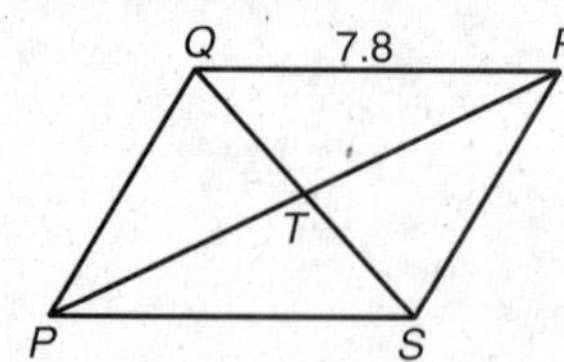

  A 6   **C** 12

  B 9.5   **D** 17.3

Holt Geometry

Problem Solving

LESSON 6-3

Conditions for Parallelograms

Use the diagram for Exercises 1 and 2.
A *pantograph* is a drawing instrument used to magnify figures.

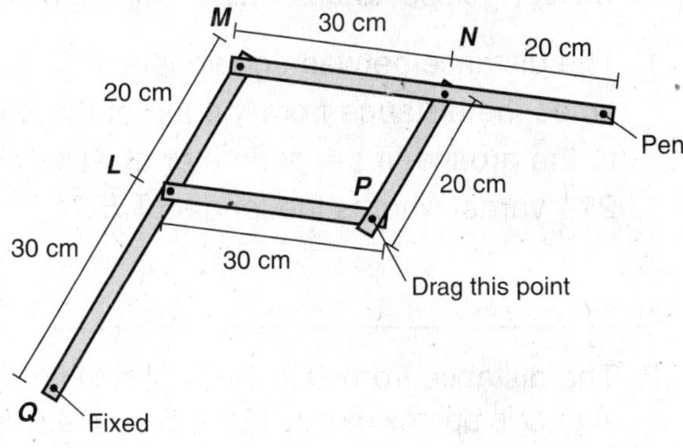

1. If you drag the point at *P* so that the angle measures change, will *LMNP* continue to be a parallelogram? Explain.

   ____

   ____

   ____

2. If you drag the point at *P* so that m∠*LMN* = 56°, what will be the measure of ∠*QLP*?

   ____

3. In the state flag of Maryland, m∠*G* = 60° and m∠*H* = 120°. Name one more condition that would allow you to conclude that *EFGH* is a parallelogram.

   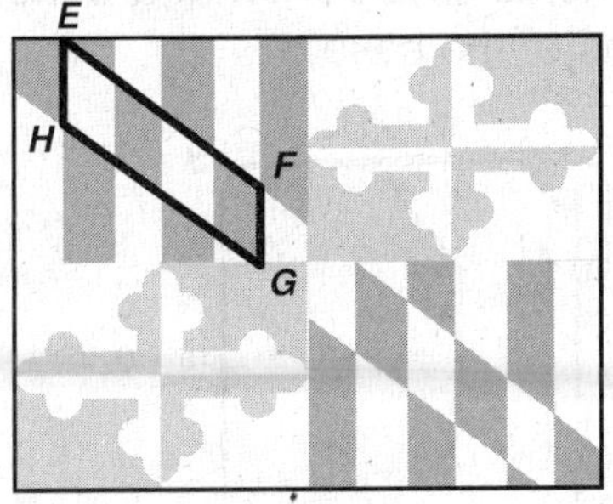

   ____

4. The graphs of $y = 2x$, $y = 2x - 5$, and $y = -x$ in the coordinate plane contain three sides of a quadrilateral. Give an equation of a line whose graph contains a segment that can complete the quadrilateral to form a parallelogram. Explain.

   ____

   ____

Choose the best answer.

5. For which value of *n* is *QRST* a parallelogram?

   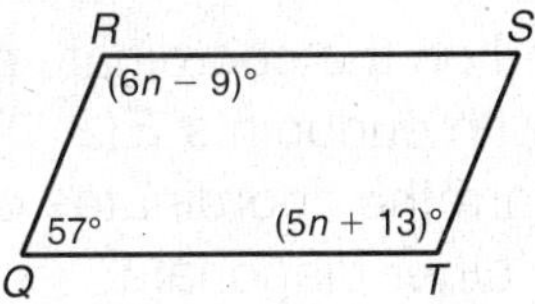

   A 15.5
   B 20.6
   C 22
   D 25

6. Under what conditions must *ABCD* be a parallelogram?

   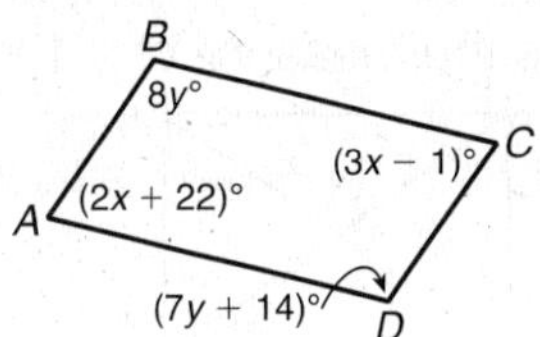

   F x = 23
   G y = 14
   H x = 23 and y = 14
   J x = 14 and y = 23

Holt Geometry

Problem Solving

Properties of Special Parallelograms

Use the diagram for Exercises 1 and 2.
The soccer goalposts determine rectangle *ABCD*.

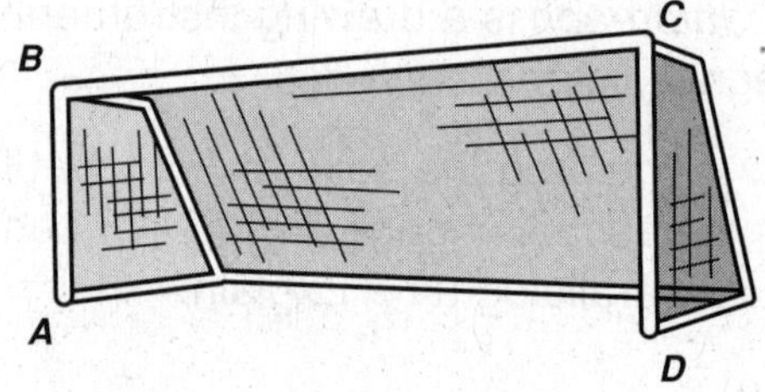

1. The distance between goalposts, *BC*, is three times the distance from the top of the goalpost to the ground. If the perimeter of *ABCD* is $21\frac{1}{3}$ yards, what is the length of $\overline{BC}$?

____

2. The distance from *B* to *D* is approximately $(x + 10)$ feet, and the distance from *A* to *C* is approximately $(2x - 5.3)$ feet. What is the approximate distance from *A* to *C*?

____

3. *MNPQ* is a rhombus. The measure of $\angle MRQ$ is $(13t - 1)°$, and the measure of $\angle PQR$ is $(7t + 4)°$. What is the measure of $\angle PQM$?

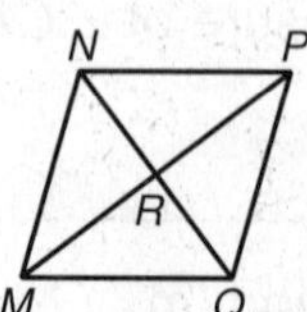

____

4. The *scissor lift* forms rhombus *PQRS* with $PQ = (7b - 5)$ meters and $QR = (2b - 0.5)$ meters. If *S* is the midpoint of $\overline{RT}$, what is the length of $\overline{RT}$?

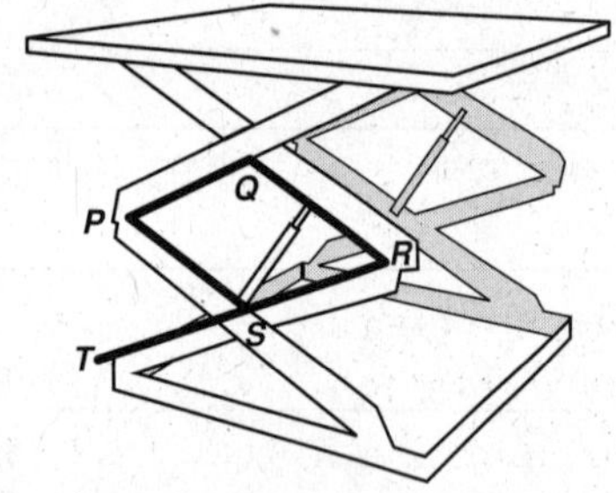

5. The diagram shows the lid of a rectangular case that holds 80 CDs. What are the dimensions of the case?

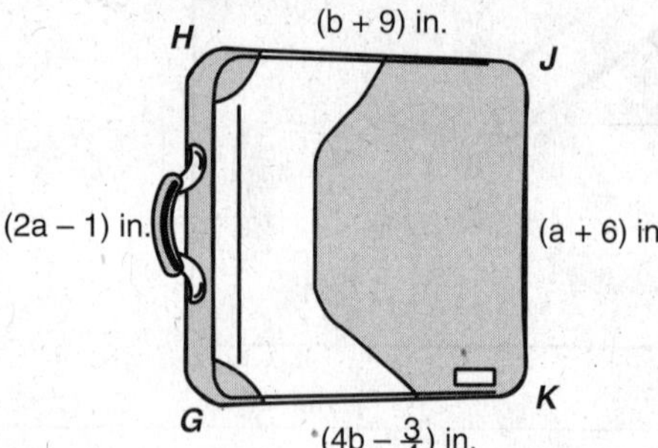

____

Choose the best answer.

6. What is the measure of $\angle 1$ in the rectangle?

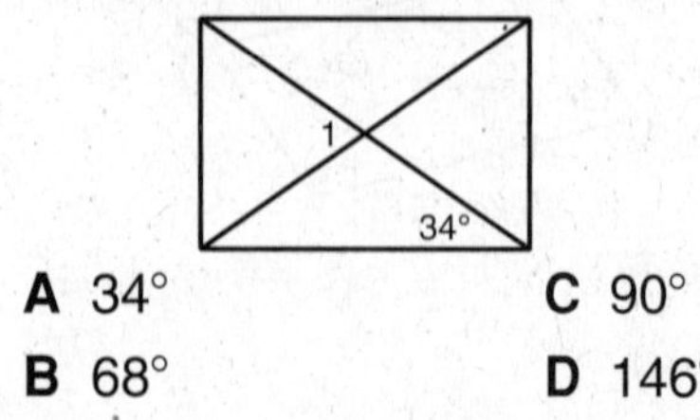

  A 34°   **C** 90°

  B 68°   **D** 146°

7. A square graphed on the coordinate plane has a diagonal with endpoints $E(2, 3)$ and $F(0, -3)$. What are the coordinates of the endpoints of the other diagonal?

  F $(4, -1)$ and $(-2, 1)$

  G $(4, 0)$ and $(-2, 1)$

  H $(4, -1)$ and $(-3, 1)$

  J $(3, -1)$ and $(-2, 1)$

Holt Geometry

Problem Solving

Conditions for Special Parallelograms

1. An amusement park has a rectangular observation deck with walkways above the bungee jumping and sky jumping. The distance from the center of the deck to points *E*, *F*, *G*, and *H* is 15 meters. Explain why *EFGH* must be a rectangle.

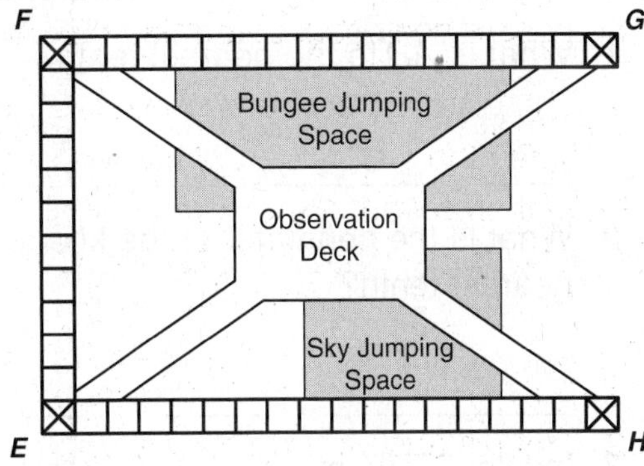

____

____

____

____

2. In the mosaic, $\overline{AB} \parallel \overline{CD}$ and $\overline{BC} \parallel \overline{DA}$. If *AB* = 4 inches and *BC* = 4 inches, can you conclude that *ABCD* is a square? Explain.

____

____

3. If $\overline{TV} \cong \overline{US}$, explain why the basketball backboard must be a rectangle.

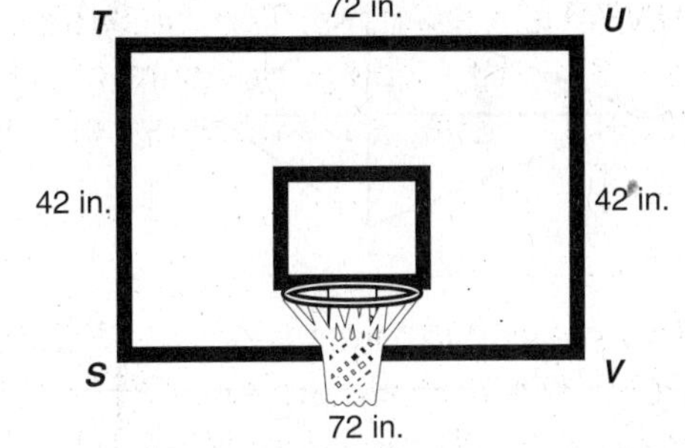

____

____

____

Choose the best answer.

4. The vertices of a parallelogram are *N*(0, −4), *P*(6, −1), *Q*(4, 3), and *R*(−2, 0). Classify the parallelogram as specifically as possible.

  A rectangle only

  B square

  C rhombus only

  D quadrilateral

5. Choose the best description for the quadrilateral.

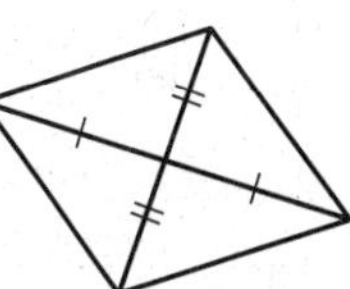

  F parallelogram

  G parallelogram and rectangle

  H parallelogram and rhombus

  J parallelogram and square

6. In parallelogram *KLMN*, m∠*L* = (4*w* + 5)°. Choose the value of *w* that makes *KLMN* a rectangle.

  A 90   **C** 43.75

  B 85   **D** 21.25

7. The coordinates of three vertices of quadrilateral *ABCD* are *A*(3, −1), *B*(10, 0), and *C*(5, 5). For which coordinates of *D* will the quadrilateral be a rhombus?

  F (−1, 4)   **H** (−1, 3)

  G (−2, 4)   **J** (−2, 3)

Holt Geometry

Problem Solving
Properties of Kites and Trapezoids

Use the figure of the kite for Exercises 1 and 2.

1. What is *AD* to the nearest tenth?

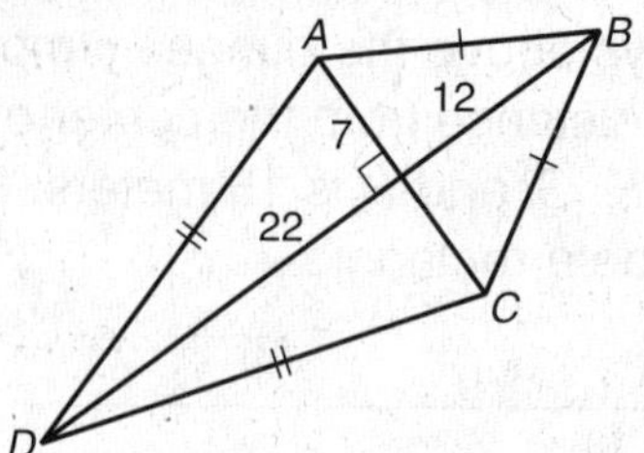

____

2. What is the perimeter of the kite to the nearest tenth?

____

3. In kite *STUV*, m∠*TUW* = 35° and m∠*WSV* = 21°. What is the measure of ∠*UVS*?

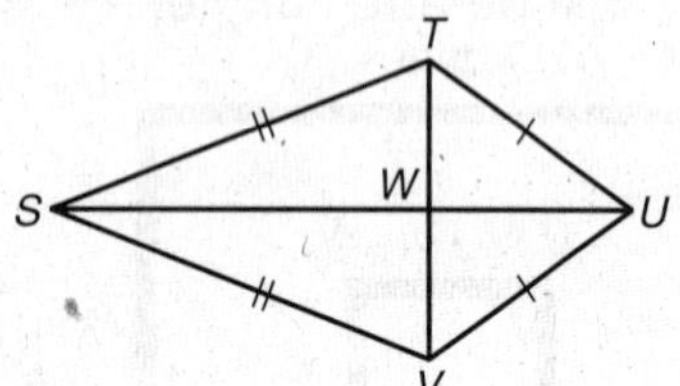

4. A car window is in the shape of a trapezoid. When the window is halfway down, the top is $\overline{KL}$, the midsegment of *FGHJ*. If *KL* = 23 inches, what is *GH*?

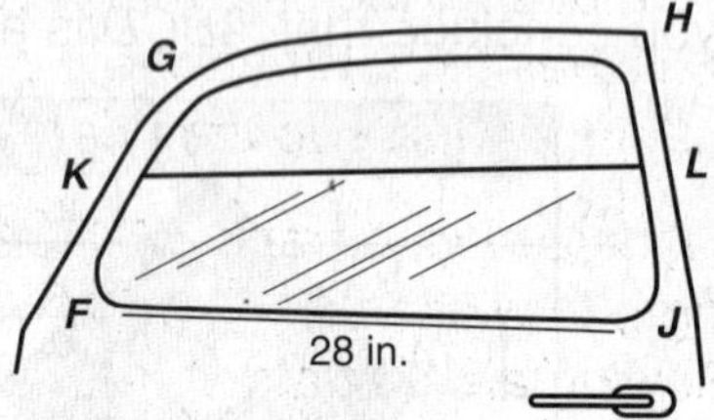

Choose the best answer.

5. Trapezoid *PQRS* has base angles that measure $(9r + 21)°$ and $(15r - 21)°$. Find the value of *r* so that *PQRS* is isosceles.

   A 3
   B 5
   C 7
   D 14

6. In kite *KLMN*, find the measure of ∠*M*.

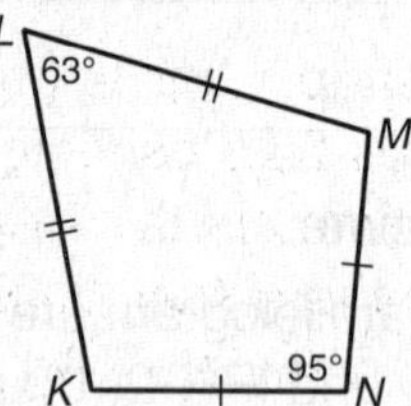

   F 100.5°   **H** 122°
   G 101°   **J** 130°

7. In the design, eight isosceles trapezoids surround a regular octagon. What is the measure of ∠*B* in trapezoid *ABCD*?

   A 35°
   B 45°
   C 55°
   D 65°

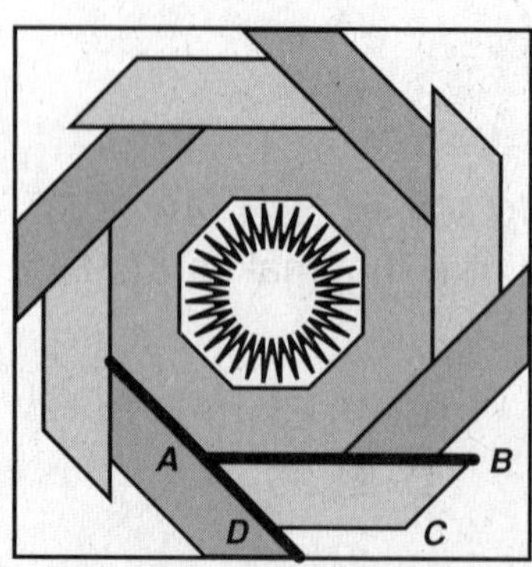

Holt Geometry

LESSON 7-1
Problem Solving
Ratio and Proportion

1. For a certain type of tropical fish, it is recommended that you have no more than 2 fish per 10 gallons of water. How many fish could you have in a fish tank that holds 35 gallons of water? _______________

2. A library is being expanded, and the new wing's length is to be 50 feet greater than its width. A diagram of the new wing is shown. What are the actual dimensions of the new wing?

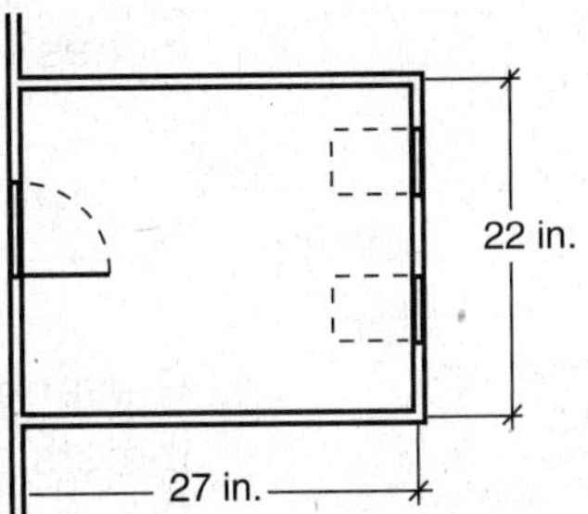

____

3. The *Titanic* was 882 feet 9 inches long. You can build a model of the ship that is 2 feet 6 inches long and 6 inches high. What was the approximate height of the *Titanic* to the nearest inch? _______________

4. The *aspect ratio,* or ratio of length to width of the viewing area, of a wide-screen 42-inch television is 16 : 9. What are the dimensions of the rectangular viewing area to the nearest tenth?

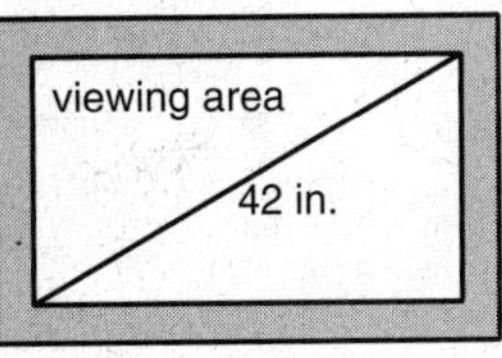

____

Choose the best answer.

5. In a museum gift shop, a miniature *Acrocanthosaurus* is 22.8 centimeters long and 15.9 centimeters tall. The package says that the actual dinosaur was approximately 9 meters long. About how tall was the dinosaur?

A 6.3 m   **C** 12.9 m

B 7.7 m   **D** 40.3 m

7. Write a ratio expressing the slope of the hypotenuse in right triangle *MNP*.

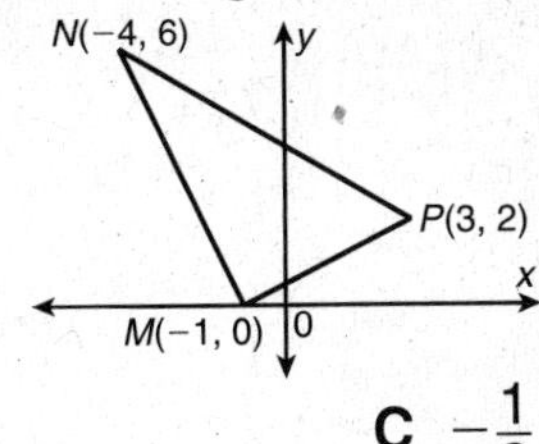

A $-\frac{7}{4}$ **C** $-\frac{1}{2}$

B $-\frac{4}{7}$ **D** $\frac{1}{4}$

6. A model airplane has a wingspan of about 15 inches. The actual airplane has a wingspan of 30 feet and a length of 42 feet. How long is the model?

F 11 in.   **H** 21 in.

G 14 in.   **J** 30 in.

8. The ratio of the interior angle measures of a pentagon is 2 : 2 : 3 : 4 : 6. What is the measure of the smallest angle to the nearest degree?

F 32°   **H** 95°

G 64°   **J** 191°

Holt Geometry

LESSON 7-2 **Problem Solving**
Ratios in Similar Polygons

1. *EFGH ~ JKLM.* What is the value of *x*?

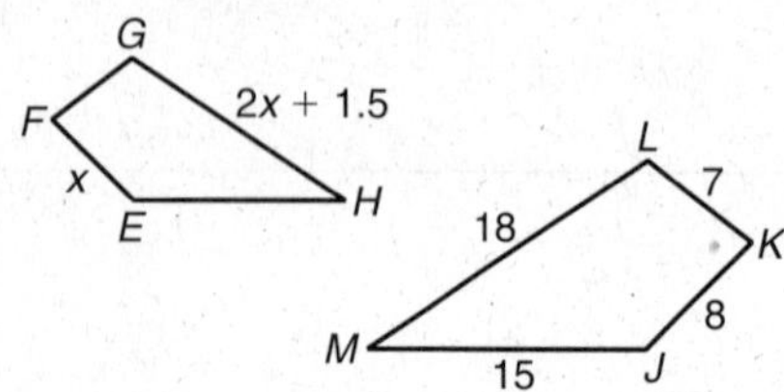

2. The ratio of a model scale die cast motorcycle is 1 : 18. The model is $5\frac{1}{4}$ inches long. What is the length of the actual motorcycle in feet and inches?

____

3. A diagram of a new competition swimming pool is shown. If the width of the pool is 25 meters, find the length of the actual pool.

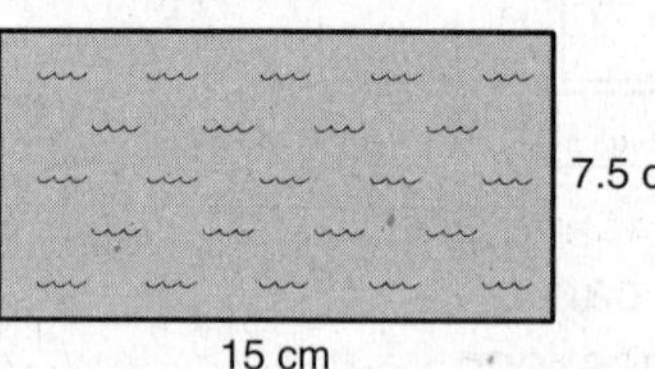

4. Rectangle A has side lengths 16.4 centimeters and 10.8 centimeters. Rectangle B has side lengths 10.25 centimeters and 6.75 centimeters. Determine whether the rectangles are similar. If so, write the similarity ratio.

Choose the best answer.

5. A pet store has various sizes of guinea pig cages. A diagram of the top view of one of the cages is shown. What are possible dimensions of this cage?

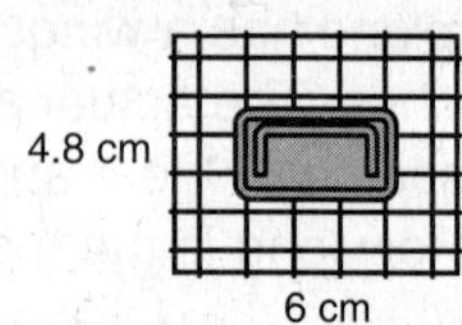

A 28 in. by 24 in.   **C** 30 in. by 24 in.
B 28 in. by 18 in.   **D** 30 in. by 18 in.

7. △*QRS ~* △*TUV.* Find the value of *y.*

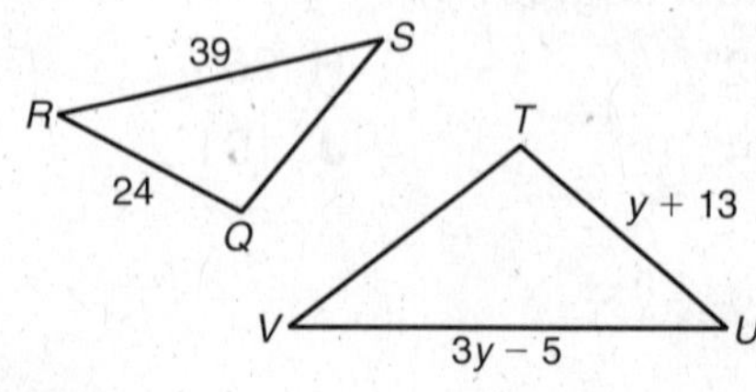

A 3.6   **C** 19
B 5.5   **D** 33

6. A gymnasium is 96 feet long and 75 feet wide. On a blueprint, the gymnasium is 5.5 inches long. To the nearest tenth of an inch, what is the width of the gymnasium on the blueprint?

F 3.7 in.   **H** 7.0 in.
G 4.3 in.   **J** 13.6 in.

8. △*ABC* has side lengths 14, 8, and 10.4. What are possible side lengths of △*DEF* if △*ABC ~* △*DEF*?

F 28, 20, 20.8
G 35, 16, 20.8
H 28, 20, 26
J 35, 20, 26

Holt Geometry

Problem Solving
Triangle Similarity: AA, SSS, and SAS

Use the diagram for Exercises 1 and 2.

In the diagram of the tandem bike, $\overline{AE} \parallel \overline{BD}$.

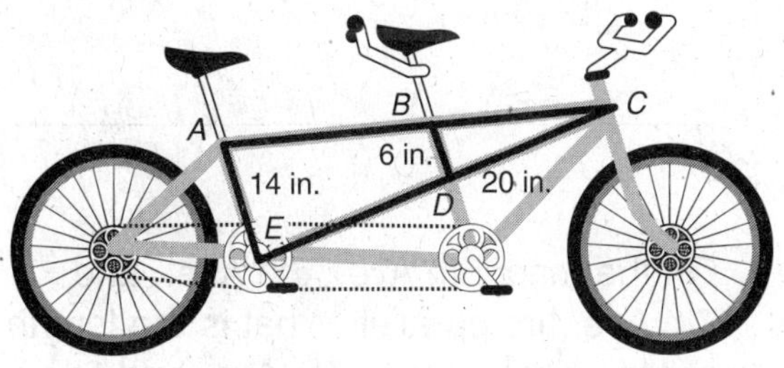

1. Explain why $\triangle CBD \sim \triangle CAE$.

____

____

____

2. Find CE to the nearest tenth. _______________________________________

3. Is $\triangle WXZ \sim \triangle XYZ$? Explain.

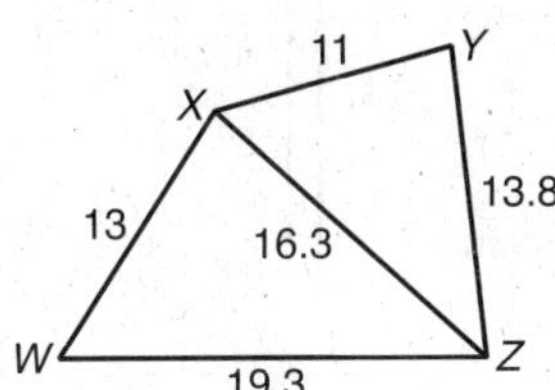

____

4. Find RQ. Explain how you found it.

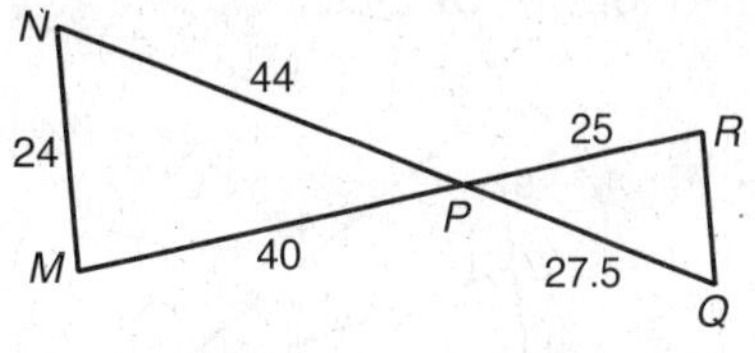

____

____

____

Choose the best answer.

5. Find the value of x that makes $\triangle FGH \sim \triangle JKL$.

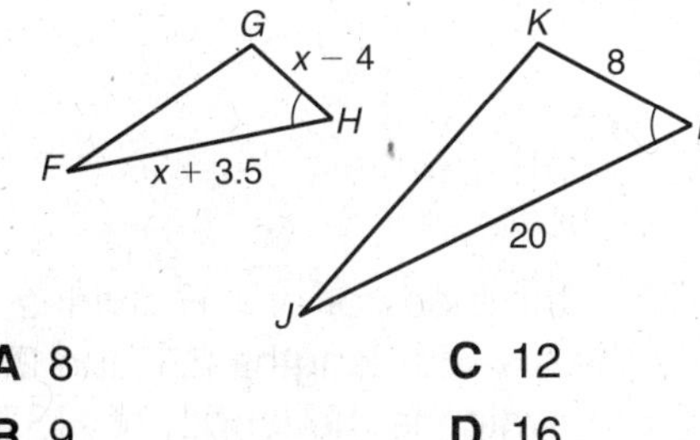

  A 8   **C** 12

  B 9   **D** 16

6. Triangle STU has vertices at $S(0, 0)$, $T(2, 6)$, and $U(8, 2)$. If $\triangle STU \sim \triangle WXY$ and the coordinates of W are $(0, 0)$, what are possible coordinates of X and Y?

  F $X(1, 3)$ and $Y(4, 1)$

  G $X(1, 3)$ and $Y(2, 0)$

  H $X(3, 1)$ and $Y(2, 4)$

  J $X(0, 3)$ and $Y(4, 0)$

7. To measure the distance EF across the lake, a surveyor at S locates points E, F, G, and H as shown. What is EF?

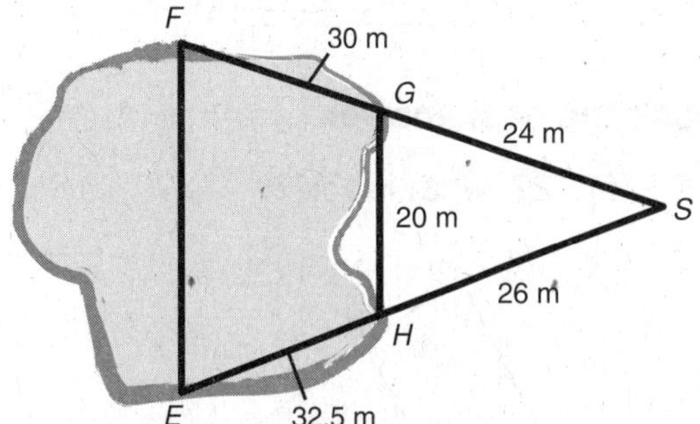

  A 25 m   **C** 45 m

  B 36 m   **D** 90 m

Holt Geometry

Problem Solving
Applying Properties of Similar Triangles

1. Is $\overline{GF} \parallel \overline{HJ}$ if $x = 5$? Explain.

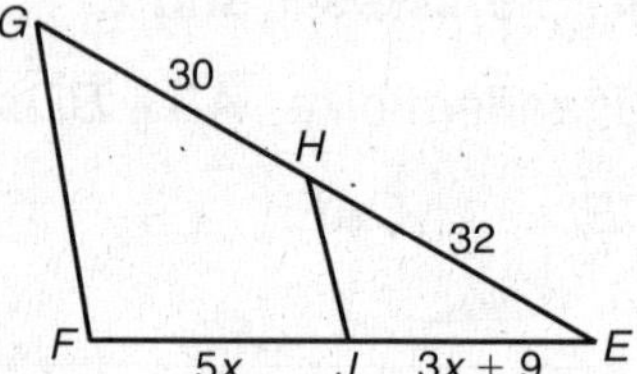

2. On the map, 5th Ave., 6th Ave., and 7th Ave. are parallel. What is the length of Main St. between 5th Ave. and 6th Ave.?

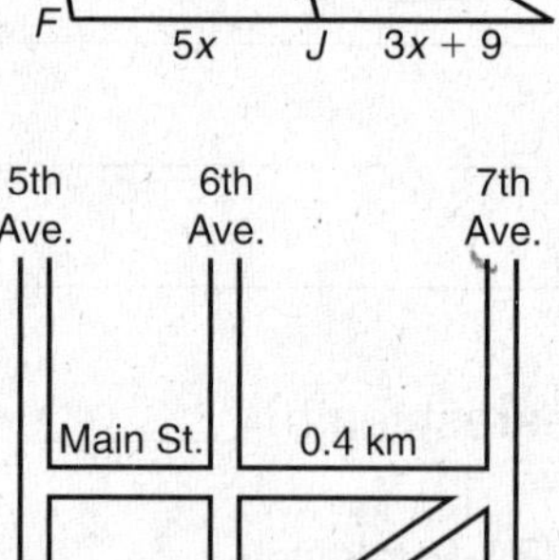

3. Find the length of $\overline{BC}$.

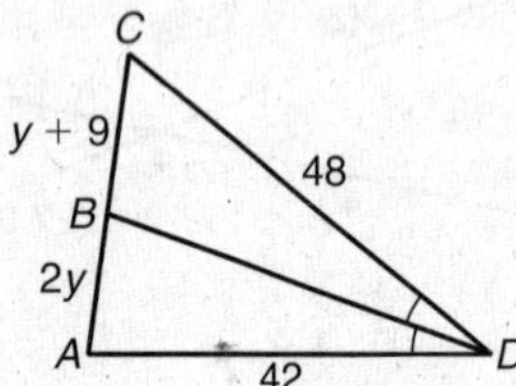

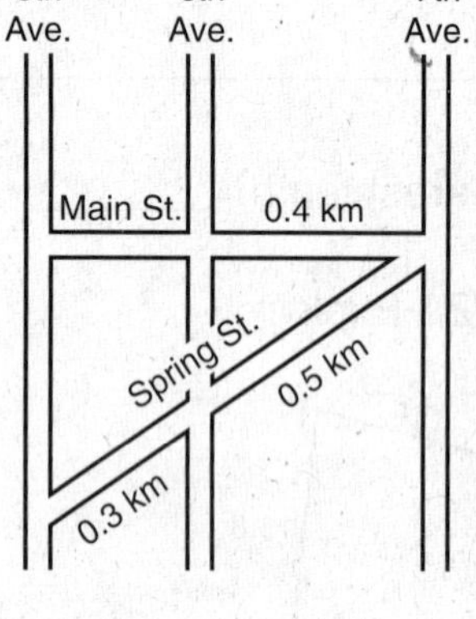

4. The figure shows three lots in a housing development. If the boundary lines separating the lots are parallel, what is GF to the nearest tenth?

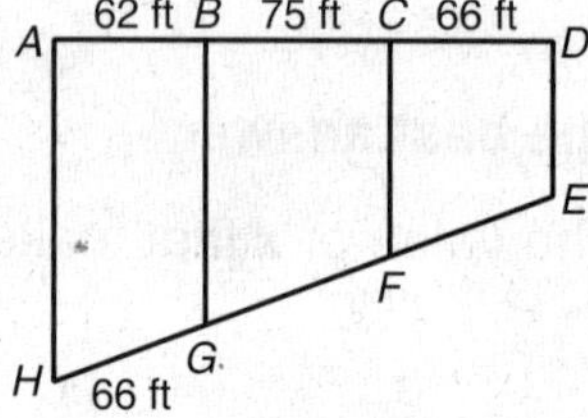

Choose the best answer.

5. If $LM = 22$, what is PM?

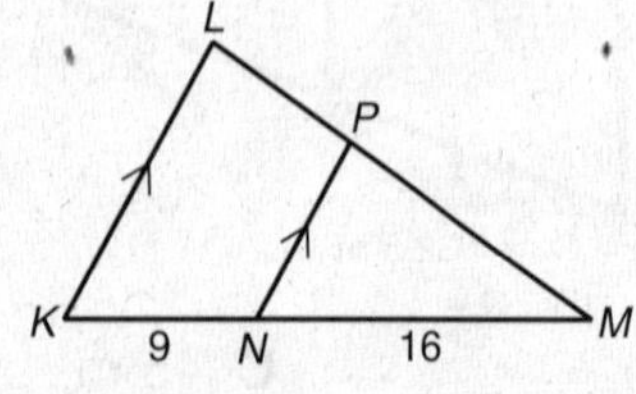

A 7.92 **C** 14.08

B 12.38 **D** 29.92

6. In $\triangle QRS$, the bisector of $\angle R$ divides $\overline{QS}$ into segments with lengths 2.1 and 2.8. If $RQ = 3$, which is the length of $\overline{RS}$?

F 2 **H** 4

G 2.25 **J** 4.5

7. In $\triangle CDE$, the bisector of $\angle C$ divides $\overline{DE}$ into segments with lengths $4x$ and $x + 13$. If $CD = 24$ and $CE = 32$, which is the length of $\overline{DE}$?

A 20 **C** 26

B 24 **D** 28

Holt Geometry

Problem Solving
Using Proportional Relationships

1. A student is standing next to a sculpture. The figure shows the shadows that they cast. What is the height of the sculpture?

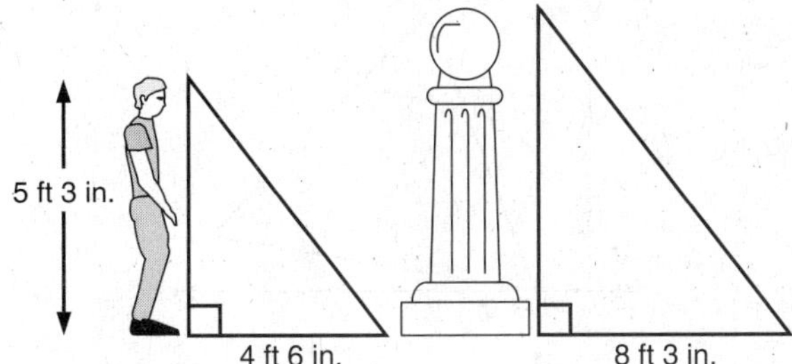

2. At the halftime show during a football game, a marching band is to form a rectangle 50 yards by 16 yards. The conductor wants to plan out the band members' positions using a 14- by 8.5-in. sheet of paper. What scale should she use to fit both dimensions of the rectangle on the page? (Use whole inches and yards.)

3. An artist makes a scale drawing of a new lion enclosure at the zoo. The scale is 1 in : 25 ft. On the drawing, the length of the enclosure is $7\frac{1}{4}$ inches. What is the actual length of the lion enclosure?

4. A room is 14 feet long and 11 feet wide. If you made a scale drawing of the top view of the room using the scale $\frac{1}{2}$ in = 2 ft, what would be the length and width of the room in your drawing?

Choose the best answer.

5. A visual-effects model maker for a movie draws a spaceship using a ratio of 1 : 24. The drawing of the spaceship is 22 inches long. What is the length of the spaceship in the movie?

   A 4 ft   **C** 44 ft

   B 8 ft   **D** 528 ft

6. A free-fall ride at an amusement park casts a shadow $43\frac{2}{3}$ feet long. At the same time, a 6-foot-tall person standing in line casts a shadow 2 feet long. What is the height of the ride?

   F $21\frac{5}{6}$ ft   **H** $98\frac{1}{4}$ ft

   G $65\frac{1}{2}$ ft   **J** 131 ft

7. The scale of the park map is 1.5 cm = 60 m. Which is the best estimate for the actual distance between the horse stables and the picnic area?

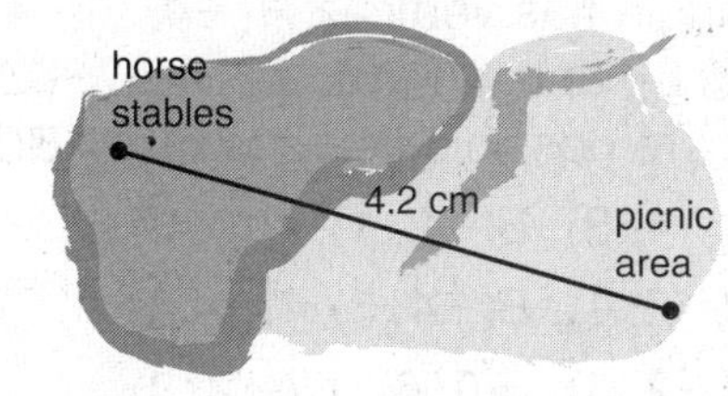

   A 21.4 m   **C** 168.0 m

   B 90.0 m   **D** 288.0 m

8. A hot-air balloon is 26.8 meters tall. Use the scale drawing to find the actual distance across the hot-air balloon.

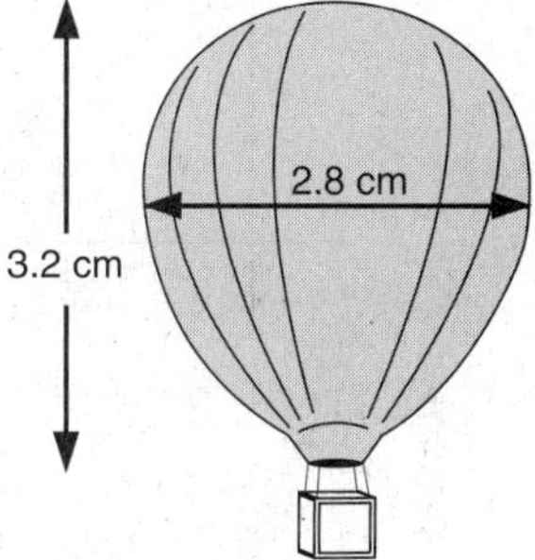

   F 23.45 m   **H** 75.0 m

   G 30.6 m   **J** 85.8 m

Holt Geometry

Problem Solving
Dilations and Similarity in the Coordinate Plane

1. The figure shows a photograph on grid paper. What are the coordinates of C' if the photograph is enlarged with scale factor $\frac{4}{3}$?

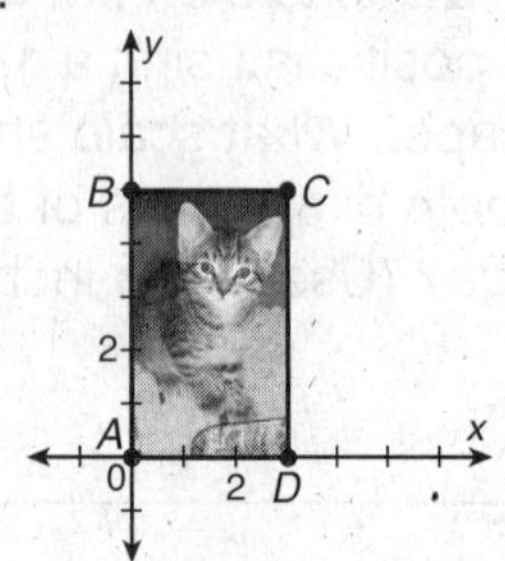

2. In the figure, $\triangle HFJ \sim \triangle EFG$. Find the coordinates of G and the scale factor.

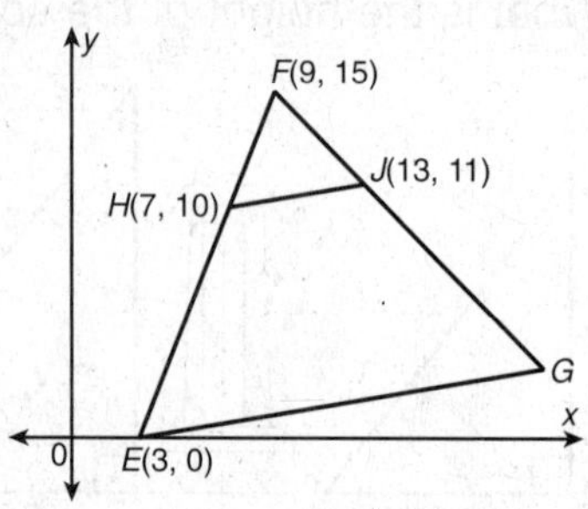

____

3. Triangle LMN has vertices $L(-10, 2)$, $M(-4, 11)$, and $N(6, -6)$. Find the vertices of the image of $\triangle LMN$ after a dilation with scale factor $\frac{5}{2}$.

____

4. Triangle HJM has vertices $H(-36, 0)$, $J(0, 20)$, and $M(0, 0)$. Triangle $H'J'M'$ has two vertices at $H'(-27, 0)$ and $M'(0, 0)$, and $\triangle H'J'M'$ is a dilation image of $\triangle HJM$. Find the coordinates of J' and the scale factor.

____

Choose the best answer.

5. The arrow is cut from a logo. The artist needs to make a copy five times as large for a sign. If the coordinates of T are $T(3, 4.5)$, what are the coordinates of T' after the arrow is dilated with scale factor 5?

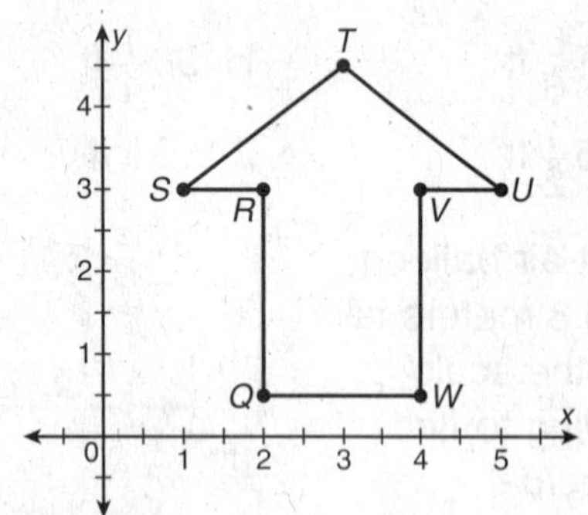

A $T'(15, 22.5)$

B $T'(7.5, 9)$

C $T'(4.5, 6.75)$

D $T'(2.5, 20)$

6. Triangle QRS has vertices $Q(-7, 3)$, $R(9, 8)$, and $S(2, 16)$. What is the scale factor if the vertices after a dilation are $Q'(-10.5, 4.5)$, $R'(13.5, 15)$, and $S'(3, 24)$?

F $\frac{1}{3}$

G $\frac{1}{2}$

H $\frac{2}{3}$

J $\frac{3}{2}$

7. A triangle has vertices $H(-4, 2)$, $J(-8, 6)$, and $K(0, 6)$. If $\triangle ABC \sim \triangle HJK$, what are possible vertices of $\triangle ABC$?

A $A(-4, 3)$, $B(-2, 1)$, $C(0, 3)$

B $A(-2, 1)$, $B(-4, 3)$, $C(0, 3)$

C $A(-2, 4)$, $B(0, 6)$, $C(-2, 8)$

D $A(-2, 4)$, $B(-8, 6)$, $C(-4, 2)$

Holt Geometry

<table><tr><td>**LESSON**
8-1</td><td></td></tr></table>

Problem Solving
Similarity in Right Triangles

1. A sculpture is 10 feet long and 6 feet wide. The artist made the sculpture so that the height is the geometric mean of the length and the width. What is the height of the sculpture to the nearest tenth of a foot?

____

3. The perimeter of $\triangle ABC$ is 56.4 cm, and the perimeter of $\triangle GHJ$ is 14.1 cm. The perimeter of $\triangle DEF$ is the geometric mean of these two perimeters. What is the perimeter of $\triangle DEF$?

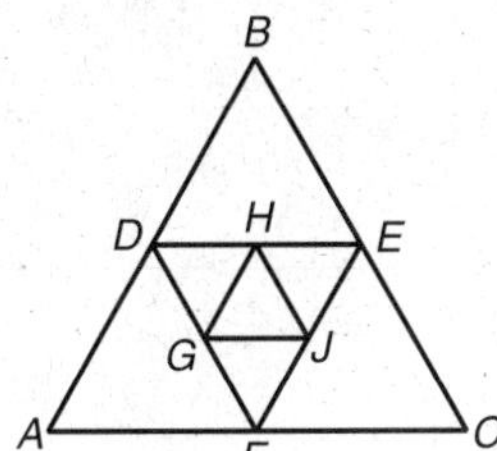

____

2. The altitude to the hypotenuse of a right triangle divides the hypotenuse into two segments that are 12 mm long and 27 mm long. What is the area of the triangle?

____

4. Tamara stands facing a painting in a museum. Her lines of sight to the top and bottom of the painting form a 90° angle. How tall is the painting?

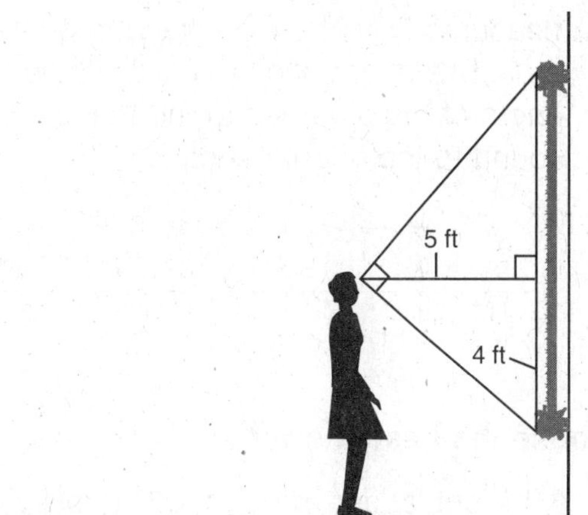

____

Choose the best answer.

5. The altitude to the hypotenuse of a right triangle divides the hypotenuse into two segments that are x cm and $4x$ cm, respectively. What is the length of the altitude?

  A $2x$ **C** $5x$

  B $2.5x$ **D** $4x^2$

7. A surveyor sketched the diagram at right to calculate the distance across a ravine. What is x, the distance across the ravine, to the nearest tenth of a meter?

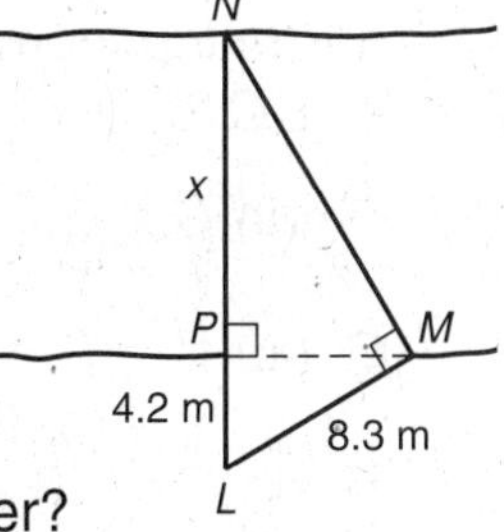

  A 7.2 m   **C** 16.4 m

  B 12.2 m   **D** 64.7 m

6. Jack stands 9 feet from the primate enclosure at the zoo. His lines of sight to the top and bottom of the enclosure form a 90° angle. When he looks straight ahead at the enclosure, the vertical distance between his line of sight and the bottom of the enclosure is 5 feet. What is the height of the enclosure?

  F 16.2 ft   **H** 23.8 ft

  G 21.2 ft   **J** 28.8 ft

Holt Geometry

Problem Solving
Trigonometric Ratios

1. A ramp is used to load a 4-wheeler onto a truck bed that is 3 feet above the ground. The angle that the ramp makes with the ground is 32°. What is the horizontal distance covered by the ramp? Round to the nearest hundredth.

____

2. Find the perimeter of the triangle. Round to the nearest hundredth.

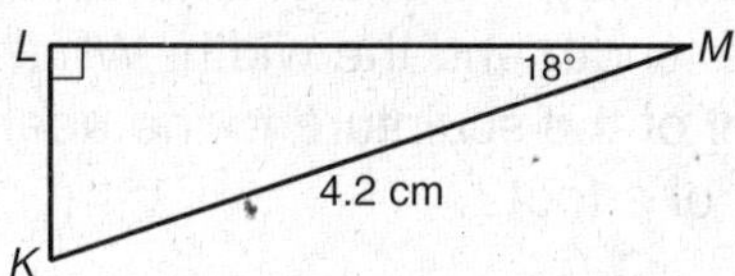

____

3. A right triangle has an angle that measures 55°. The leg adjacent to this angle has a length of 43 cm. What is the length of the other leg of the triangle? Round to the nearest tenth.

____

4. The hypotenuse of a right triangle measures 9 inches, and one of the acute angles measures 36°. What is the area of the triangle? Round to the nearest square inch.

____

Choose the best answer.

5. A 14-foot ladder makes a 62° angle with the ground. To the nearest foot, how far up the house does the ladder reach?

A 6 ft

B 7 ft

C 12 ft

D 16 ft

7. What is EF, the measure of the longest side of the sail on the model? Round to the nearest inch.

A 31 in.

B 35 in.

C 40 in.

D 60 in.

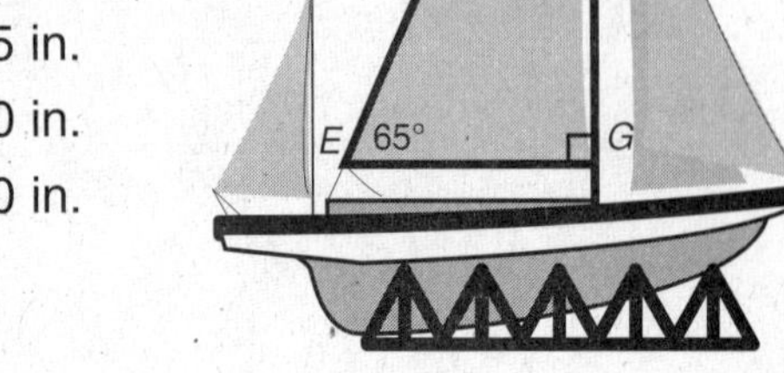

6. To the nearest inch, what is the length of the springboard shown below?

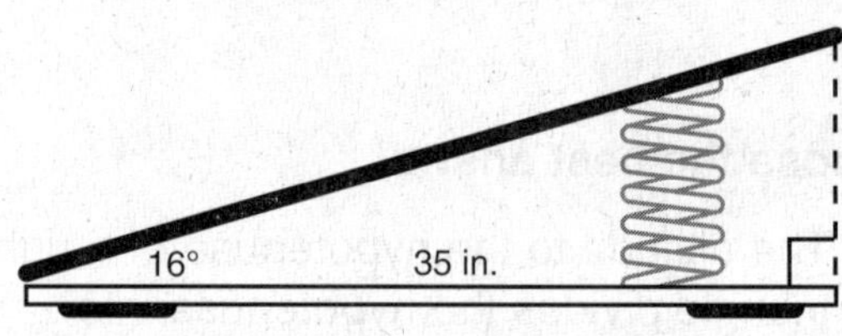

F 24 in.

G 36 in.

H 38 in.

J 127 in.

8. Right triangle ABC is graphed on the coordinate plane and has vertices at $A(-1, 3)$, $B(0, 5)$, and $C(4, 3)$. Which is a true trigonometric ratio for the measure of $\angle C$?

F $\tan C = \dfrac{1}{2}$

G $\tan C = 2$

H $\sin C = \dfrac{\sqrt{5}}{3}$

J $\cos C = 2$

Holt Geometry

Problem Solving
Solving Right Triangles

1. A road has a grade of 28.4%. This means that the road rises 28.4 ft over a horizontal distance of 100 ft. What angle does the hill make with a horizontal line? Round to the nearest degree.

____

2. Pet ramps for loading larger dogs into vehicles usually have slopes between $\frac{2}{5}$ and $\frac{1}{2}$. What is the range of angle measures that most pet ramps make with a horizontal line? Round to the nearest degree.

____

Use the side view of a water slide for Exercises 3 and 4.

The ladder, represented by $\overline{AB}$, is 17 feet long.

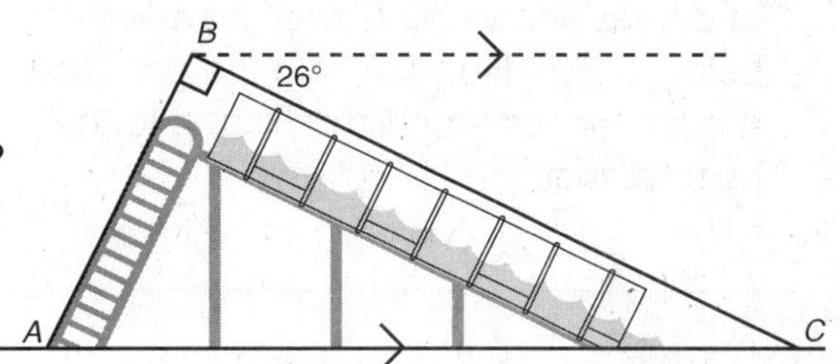

3. What is the measure of angle A, the angle that the ladder makes with a horizontal line?

____

4. What is BC, the length of the slide? Round to the nearest tenth of a foot.

____

Choose the best answer.

5. Janelle sets her treadmill grade to 6%. What is the angle that the treadmill surface makes with a horizontal line? Round to the nearest degree.

  A 3°   **C** 12°
  B 4°   **D** 31°

6. The coordinates of the vertices of $\triangle RST$ are $R(3, 3)$, $S(8, 3)$, and $T(8, -6)$. What is the measures of angle T? Round to the nearest degree.

  F 18°   **H** 61°
  G 29°   **J** 65°

7. If $\cos A = 0.28$, which angle in the triangles below is $\angle A$?

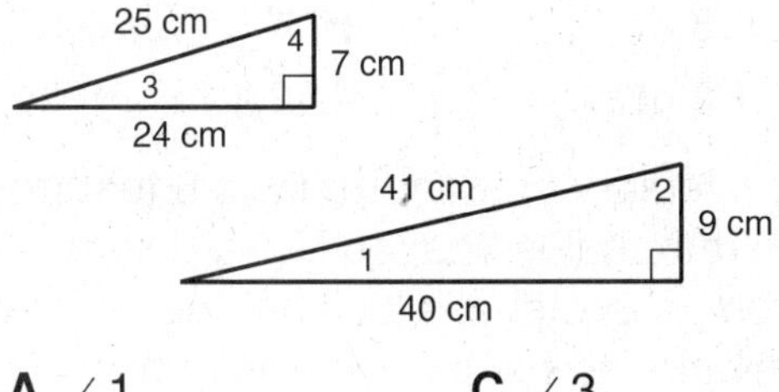

  A $\angle 1$ **C** $\angle 3$
  B $\angle 2$ **D** $\angle 4$

8. Find the measure of the acute angle formed by the graph of $y = \frac{3}{4}x$ and the x-axis. Round to the nearest degree.

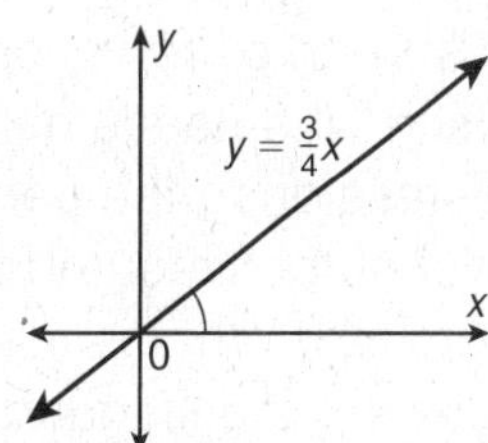

  F 37°   **H** 49°
  G 41°   **J** 53°

Holt Geometry

<table><tr><td>**LESSON**
8-4</td><td></td></tr></table>

Problem Solving
Angles of Elevation and Depression

1. Mayuko is sitting 30 feet high in a football stadium. The angle of depression to the center of the field is 14°. What is the horizontal distance between Mayuko and the center of the field? Round to the nearest foot.

2. A surveyor 50 meters from the base of a cliff measures the angle of elevation to the top of the cliff as 72°. What is the height of the cliff? Round to the nearest meter.

3. Shane is 61 feet high on a ride at an amusement park. The angle of depression to the park entrance is 42°, and the angle of depression to his friends standing below is 80°. How far from the entrance are his friends standing? Round to the nearest foot.

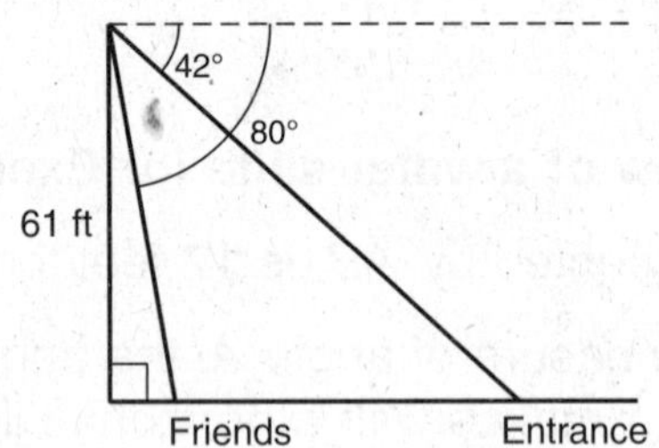

Choose the best answer.

4. The figure shows a person parasailing. What is *x*, the height of the parasailer, to the nearest foot?

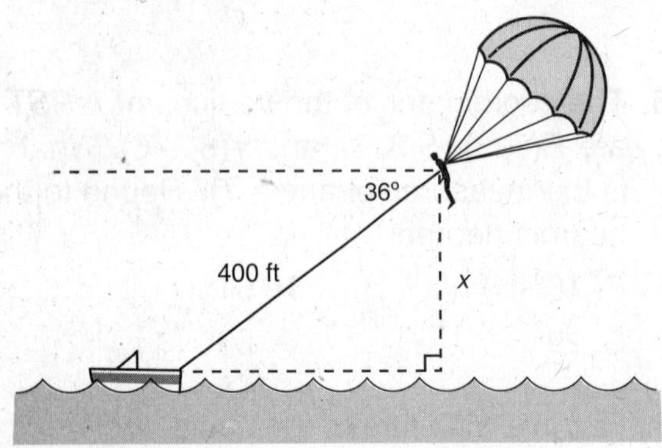

  A 235 ft   **C** 290 ft

  B 245 ft   **D** 323 ft

5. The elevation angle from the ground to the object to which the satellite dish is pointed is 32°. If *x* = 2.5 meters, which is the best estimate for *y*, the height of the satellite stand?

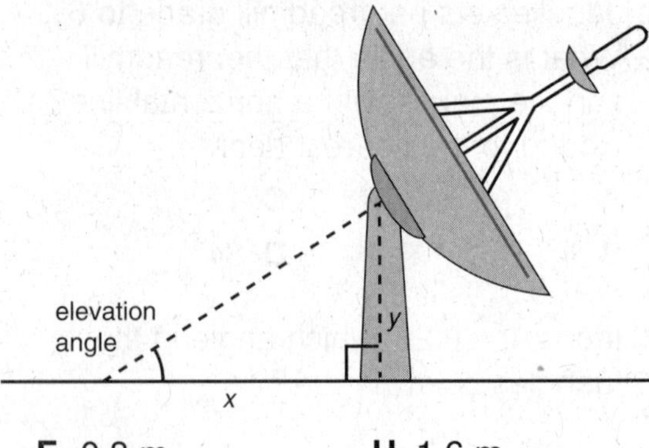

  F 0.8 m   **H** 1.6 m

  G 1.3 m   **J** 2.1 m

6. A lifeguard is in an observation chair and spots a person who needs help. The angle of depression to the person is 22°. The eye level of the lifeguard is 10 feet above the ground. What is the horizontal distance between the lifeguard and the person? Round to the nearest foot.

  A 4 ft   **C** 25 ft

  B 11 ft   **D** 27 ft

7. At a topiary garden, Emily is 8 feet from a shrub that is shaped like a dolphin. From where she is looking, the angle of elevation to the top of the shrub is 46°. If she is 5 feet tall, which is the best estimate for the height of the shrub?

  F 6 ft   **H** 10 ft

  G 8 ft   **J** 13 ft

Holt Geometry

Problem Solving
Law of Sines and Law of Cosines

1. The map shows three earthquake centers for one week in California. How far apart were the earthquake centers at points *A* and *C*? Round to the nearest tenth.

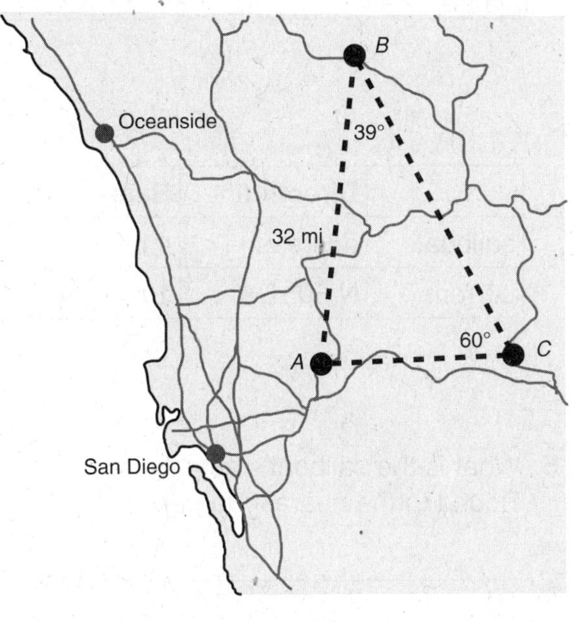

2. A BMX track has a starting hill as shown in the diagram. What is the length of the hill, *WY*? Round to the nearest tenth.

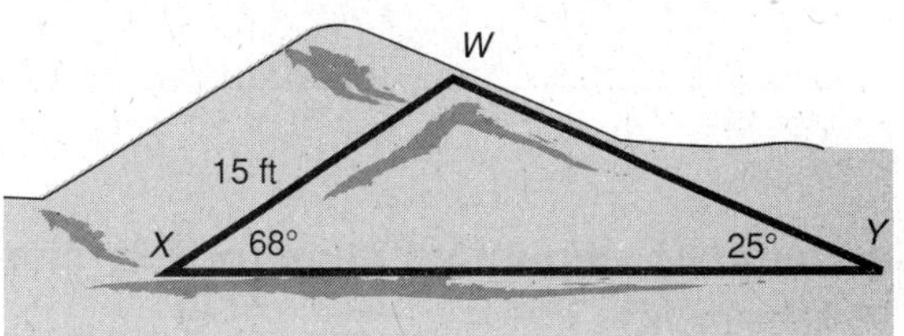

3. The edges of a triangular cushion measure 8 inches, 3 inches, and 6 inches. What is the measure of the largest angle of the cushion to the nearest degree?

4. The coordinates of the vertices of $\triangle HJK$ are $H(0, 4)$, $J(5, 7)$, and $K(9, -1)$. Find the measure of $\angle H$ to the nearest degree.

Choose the best answer. Use the following information and diagram for Exercises 5 and 6.

To find the distance across a bay, a surveyor locates points *Q*, *R*, and *S* as shown.

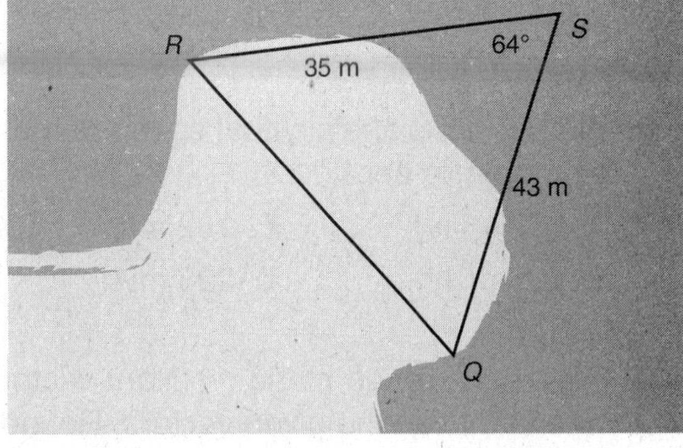

5. What is *QR* to the nearest tenth?

  A 8 m   **C** 41.9 m

  B 35.2 m   **D** 55.4 m

6. What is $m\angle Q$ to the nearest degree?

  F 43°   **H** 67°

  G 49°   **J** 107°

7. Two angles of a triangle measure 56° and 77°. The side opposite the 56° angle is 29 cm long. What is the measure of the shortest side? Round to the nearest tenth.

  A 23.4 cm   **C** 32.9 cm

  B 25.6 cm   **D** 34.1 cm

8. Which is the best estimate for the perimeter of a triangle if two sides measure 7 inches and 10 inches, and the included angle between the two sides is 82°?

  F 11.4 in.   **H** 28.4 in.

  G 12.2 in.   **J** 39.9 in.

Holt Geometry

Problem Solving
Vectors

1. The velocity of a wave is given by the vector $\langle 7, 3 \rangle$. Find the direction of the vector to the nearest degree.

2. Hikers set out on a course given by the vector $\langle 6, 11 \rangle$. What is the length of the trip to the nearest unit?

Use the following information for Exercises 3–5.

A sailboat is traveling in water with a current shown in the table.

	Direction	Rate
sailboat	due east	4 mi/h
current	N 60° E	1 mi/h

3. What is the resultant vector in component form? Round to the nearest tenth.

4. What is the sailboat's actual speed to the nearest tenth?

5. What is the sailboat's actual direction? Round to the nearest degree.

Choose the best answer.Use the following information for Exercises 6 and 7.

A small plane is flying with the conditions shown in the table.

	Direction	Rate
plane	due north	200 mi/h
wind	due east	28 mi/h

6. What is the plane's actual speed to the nearest mile per hour?

   A 172 mi/h   C 202 mi/h

   B 198 mi/h   D 228 mi/h

7. What is the direction of the plane to the nearest degree?

   F 82°   H 16°

   G 41°   J 8°

8. Find the direction of the resultant vector when you add the given vectors. Round to the nearest degree.

$$\vec{u} = \langle -4, 3 \rangle \text{ and } \vec{v} = \langle 1, 3 \rangle$$

   A N 63° E   C N 27° W

   B N 63° W   D N 27° E

9. A person in a canoe leaves shore at a bearing of N 45° W and paddles at a constant speed of 2 mi/h. There is a 1.5 mi/h current moving due west. What is the canoe's actual speed?

   F 0.5 mi/h   H 3.2 mi/h

   G 0.8 mi/h   J 3.5 mi/h

Holt Geometry

LESSON 9-1 Problem Solving
Developing Formulas for Triangles and Quadrilaterals

1. The area of trapezoid *HJKL* is 385 mm^2. Find *LM* to the nearest tenth.

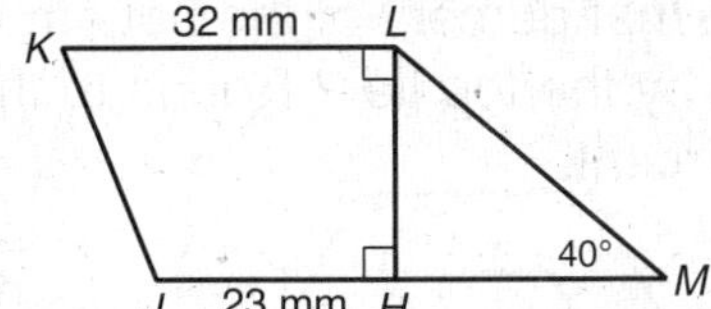

2. Samantha is buying stones for a 16 foot by 3 foot walkway. She needs to buy 10% extra for cutting stones for the corners and ends. The stones are rectangles 7 inches long and 4 inches wide and cost $0.97 each. About how much will the stones cost for her walkway?

3. The length of a rectangular pool is 2 feet less than twice the width. If the area of the pool is 264 ft^2, what are the dimensions of the pool?

4. Find the area of the kite. Round to the nearest tenth.

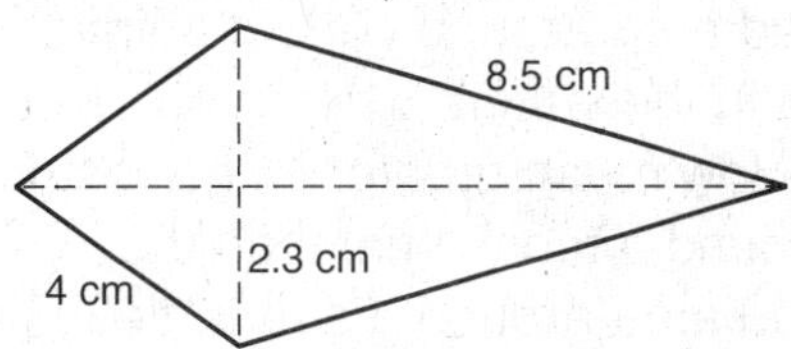

Choose the best answer.

5. A parallelogram has sides of length 30 centimeters and 18 centimeters. One of its angles measures 58°. Which is the best estimate for the area of the parallelogram?

A 274.8 cm^2

B 286.2 cm^2

C 457.9 cm^2

D 540.0 cm^2

6. Jamie is cutting out 32 right triangles from fabric for her quilt. The shortest side of each triangle is 2 inches, and the longest side is 5 inches. How much fabric will she use to cut out all the triangles?

F 146.6 in^2 **H** 366.6 in^2

G 293.3 in^2 **J** 672 in^2

7. In rhombus *ABCD*, the length of diagonal $\overline{BD}$ is $\frac{2}{3}$ the length of diagonal $\overline{AC}$. If the area of the figure is 75 cm^2, find *BD*.

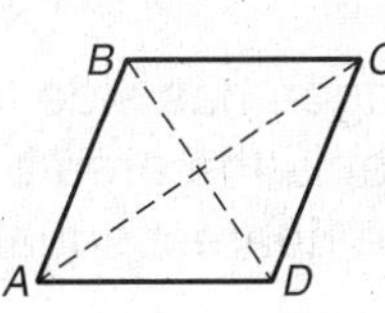

A 7.1 cm   **C** 10.6 cm

B 10 cm   **D** 15 cm

8. The house is made from 7 puzzle pieces. The pieces are triangles and parallelograms. If the area of the chimney is $\frac{1}{8}$ in^2, what is the area of $\triangle LMN$?

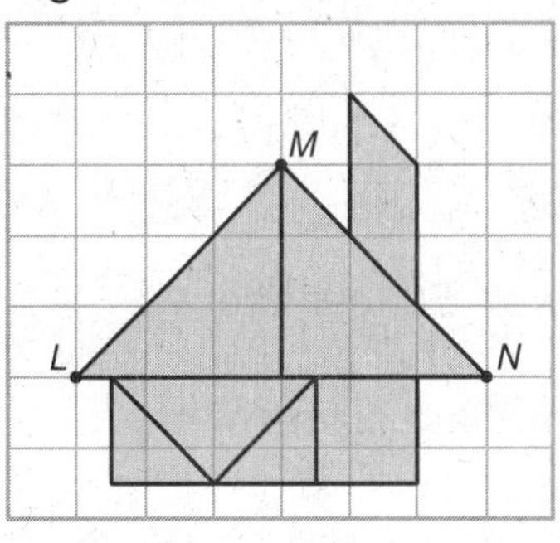

F $\frac{9}{32}$ in^2 **H** $\frac{9}{2}$ in^2

G $\frac{9}{16}$ in^2 **J** 9 in^2

Holt Geometry

Problem Solving
Developing Formulas for Circles and Regular Polygons

1. What is the area of the regular nonagon? Round to the nearest tenth.

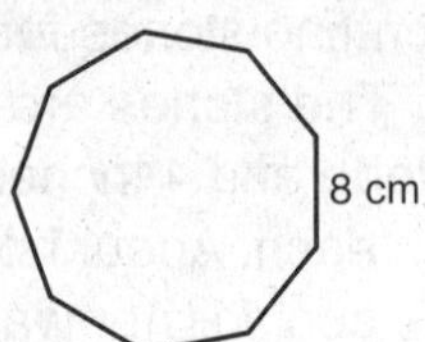

2. The top view of a two-tiered wedding cake is shown. Each tier is a regular hexagon. What percent of the bottom tier is covered by the top tier? Round to the nearest percent.

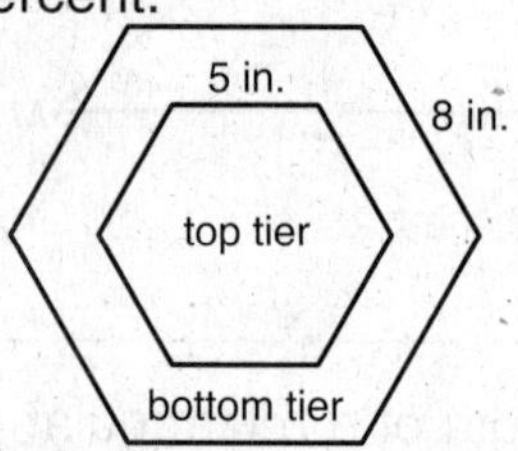

3. When diving and snorkeling, you should leave a "radius of approach," or a restricted area around certain animals that live in the waters where you are diving. How much greater is the restricted area around a monk seal than the restricted area around a sea turtle? Give your answer in terms of π.

Animal	Radius of Approach
sea turtle	20 ft
monk seal	100 ft

4. A yield sign is a regular triangle and is available in two sizes: 30 inches or 36 inches. Find how much more metal is needed to make a 36 inch sign than a 30 inch sign. Answer to the nearest percent.

Choose the best answer.

5. A regular hexagon has an apothem of 4.6 centimeters. Which is the best estimate for the area of the hexagon?

A 36.7 cm^2

B 63.5 cm^2

C 73.3 cm^2

D 146.6 cm^2

6. An amusement park ride is made up of a large circular frame that holds 50 riders. The circumference of the frame is about 138 feet. What is the diameter of the ride to the nearest foot?

F 22 ft   **H** 69 ft

G 44 ft   **J** 138 ft

7. A cyclist travels 50 feet after 7.34 rotations of her bicycle wheels. What is the approximate diameter of the wheels?

A 13 in.   **C** 26 in.

B 24 in.   **D** 28 in.

8. A regular pentagon has side length 16 inches. What is the area of the pentagon to the nearest square inch?

F 440 in^2 **H** 544 in^2

G 369 in^2 **J** 881 in^2

Holt Geometry

LESSON 9-3

Problem Solving
Composite Figures

1. Find the shaded area. Round to the nearest tenth.

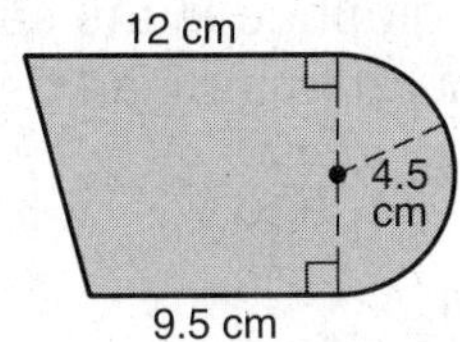

2. Jessica is painting a bedroom wall shown by the shaded area below. The cost of paint is $6.90 per quart, and each quart covers 65 square feet. What is the total cost of the paint if she applies two coats of paint to the wall?

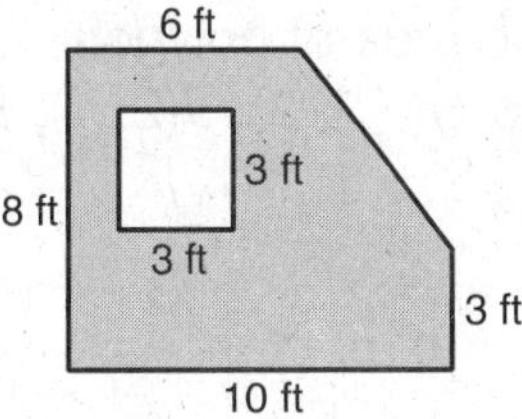

Choose the best answer.

3. Enchanted Rock State Natural Area in Fredericksburg, Texas, has a primitive camping area called Moss Lake. Which is the best estimate for this area if the length of each grid square is 10 meters?

A 1600 m^2

B 3200 m^2

C 6400 m^2

D 8000 m^2

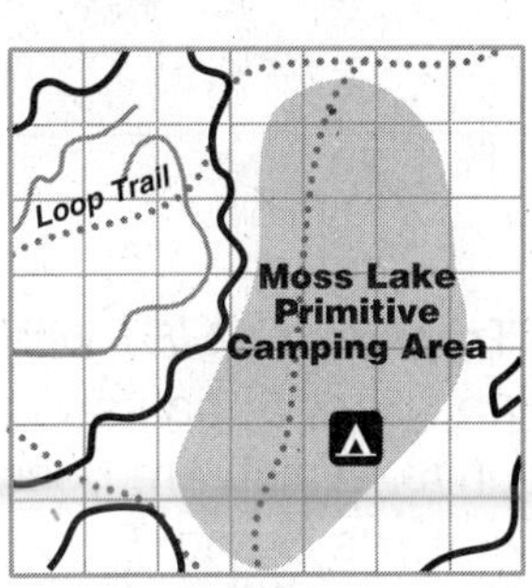

4. Find the area of the section of basketball court that is shown. Round to the nearest tenth.

F 612.7 ft^2

G 820.1 ft^2

H 1225.4 ft^2

J 2450.8 ft^2

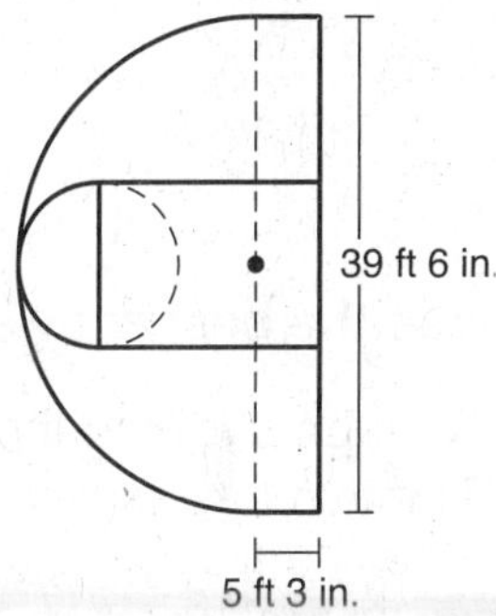

5. Find the shaded area. Round to the nearest tenth.

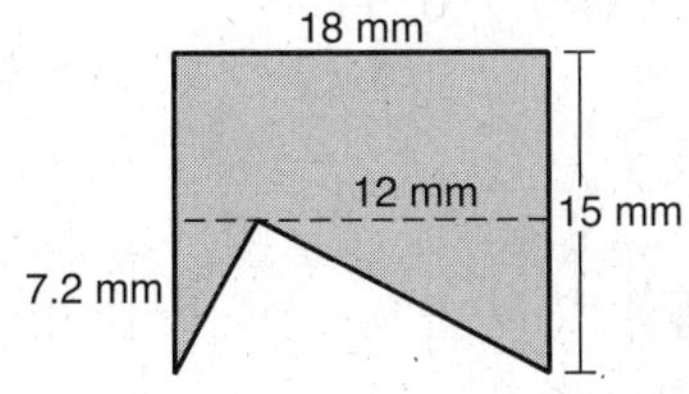

A 183.6 mm^2

B 194.4 mm^2

C 205.2 mm^2

D 216.0 mm^2

6. Which is the best estimate for the area of the pond? Each grid square represents 4 square feet.

F 24 ft^2

G 48 ft^2

H 96 ft^2

J 120 ft

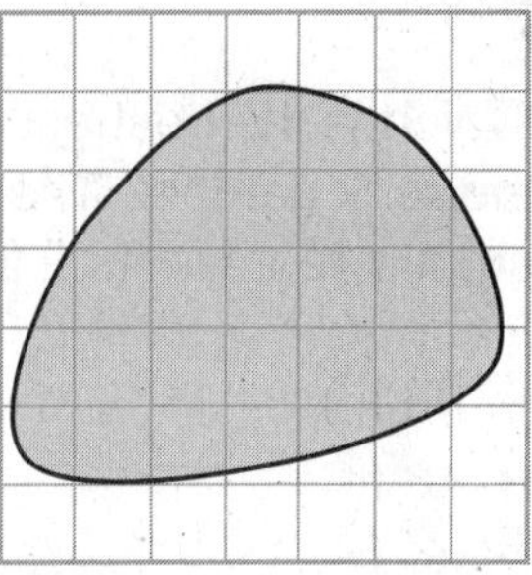

Holt Geometry

Problem Solving
Perimeter and Area in the Coordinate Plane

1. Find the perimeter and area of a polygon with vertices $A(-3, -2)$, $B(2, 4)$, $C(5, 2)$, and $D(0, -4)$. Round to the nearest tenth.

____

3. Find the area of polygon *HJKL* with vertices $H(-3, 3)$, $J(2, 1)$, $K(4, -4)$, and $L(-3, -3)$.

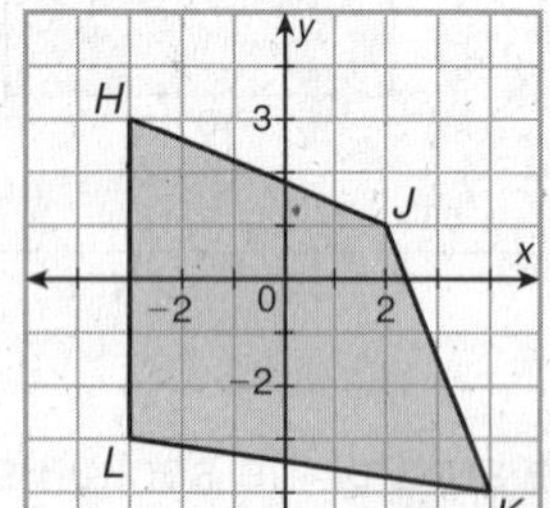

2. What are the perimeter and area of the triangle that is formed when the lines below are graphed in the coordinate plane? Round to the nearest tenth.

$y = 2x$, $y = 4$, and $y = x + 4$

____

4. The diagram represents train tracks in the children's area of a zoo. Estimate the area enclosed by the tracks. The side length of each square represents 1 meter.

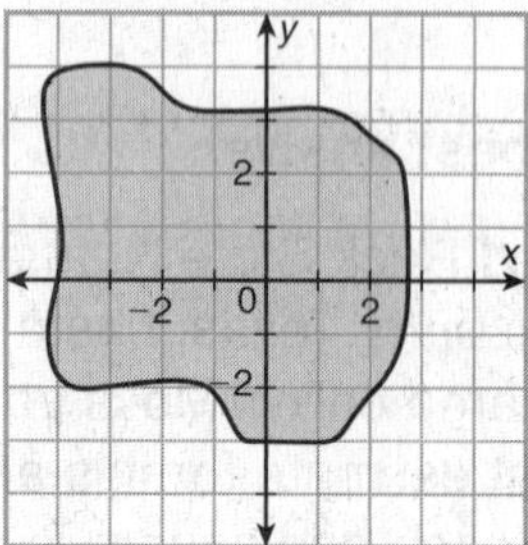

____

Choose the best answer.

5. A graph showing the top view of a circular fountain has its center at (4, 6). The circle representing the fountain passes through (2, 1). What is the area of the space covered by the fountain?

  A $\sqrt{29}\,\pi$

  B $2\sqrt{29}\,\pi$

  C 29π

  D 58π

6. Trapezoid *QRST* with vertices $Q(1, 5)$ and $R(9, 5)$ has an area of 12 square units. Which are possible locations for vertices *S* and *T*?

  F $S(6, 7)$ and $T(2, 7)$

  G $S(4, 7)$ and $T(2, 7)$

  H $S(6, 8)$ and $T(3, 8)$

  J $S(6, 1)$ and $T(3, 1)$

7. Which is the best estimate for the area of the rock garden? The side length of each square represents 2 feet.

  A 23 ft^2

  B 46 ft^2

  C 69 ft^2

  D 92 ft^2

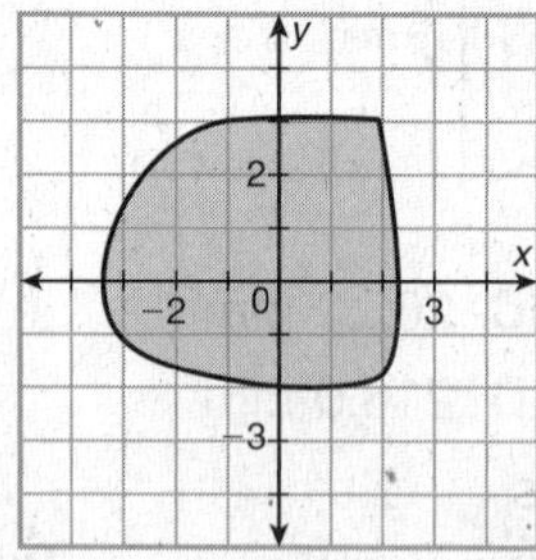

Holt Geometry

LESSON 9-5 Problem Solving
Effects of Changing Dimensions Proportionally

1. Mara has a photograph 5 inches by 7 inches. She wants to enlarge the photo so that the length and width are each tripled. Describe how the area of the photo will change.

____

2. On a map, 1 inch = 2 miles. On the map, the area of a wildlife preserve is about 3 square inches. Estimate the actual area of the preserve in acres. (*Hint:* 1 square mile = 640 acres)

____

3. A triangle has vertices $N(3, 5)$, $P(7, 2)$, and $Q(3, 1)$. Point P is moved to be twice as far from $\overline{NQ}$ as in the original triangle. Describe the effect on the area.

____

4. The length of each base of a trapezoid is divided by 2. How does the area change?

____

Use the information below for Exercises 5 and 6.

Steven's dog is on a chain 6 feet long with one end of the chain attached to the ground as shown in the diagram. Steven replaces the chain with one that is $1\frac{1}{2}$ times as long.

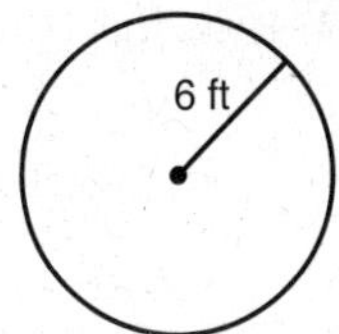

5. Describe how the circumference of the circle determined by the chain is changed.

____

6. Describe how the area of the circle determined by the chain is changed.

____

Choose the best answer.

7. In kite $RSTU$, $RT = 2.5$ centimeters and $SU = 4.3$ centimeters. Both diagonals of the kite are doubled. What happens to the area of the kite?

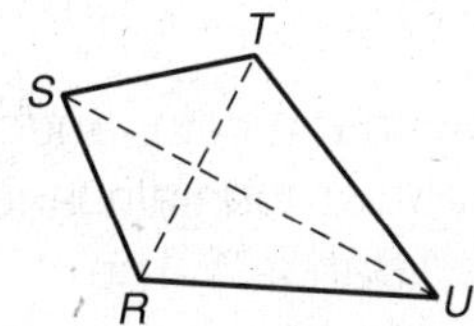

A The area is doubled.

B The area is tripled.

C The area is 4 times as great.

D The area is 8 times as great.

8. The side length of the regular hexagon is divided by 3. Which is a true statement?

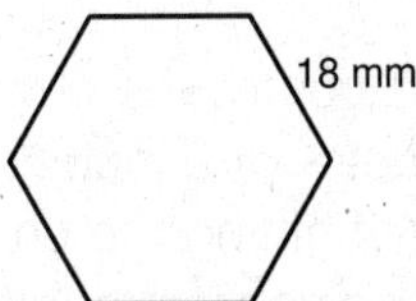

F The perimeter is divided by 9, and the area is divided by 3.

G The perimeter is divided by 3, and the area is divided by 9.

H The perimeter and area are both divided by 3.

J The perimeter and area are both divided by 9.

Holt Geometry

Problem Solving
Geometric Probability

Use the diagram of a spinner for Exercises 1 and 2.

1. Find the probability of the pointer landing on the 120° section.

____

2. Find the probability of the pointer landing on the 100° section.

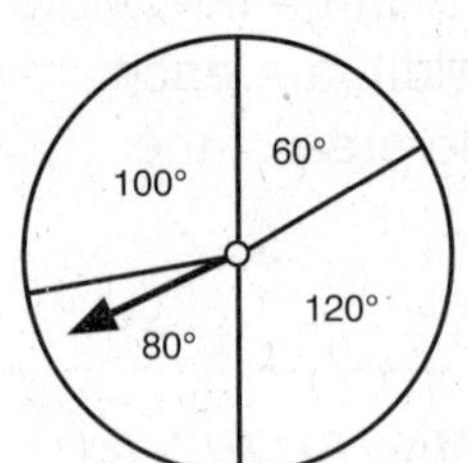

____

3. Between 4:00 P.M. and 6:30 P.M., a radio station gives a traffic report every 20 minutes. This report lasts 15 seconds. Suppose you turn on the radio between 4:00 P.M. and 6:30 P.M. Find the probability that a traffic report will be on.

4. Find the probability that a point chosen randomly inside the rectangle is in the triangle. Round to the nearest hundredth.

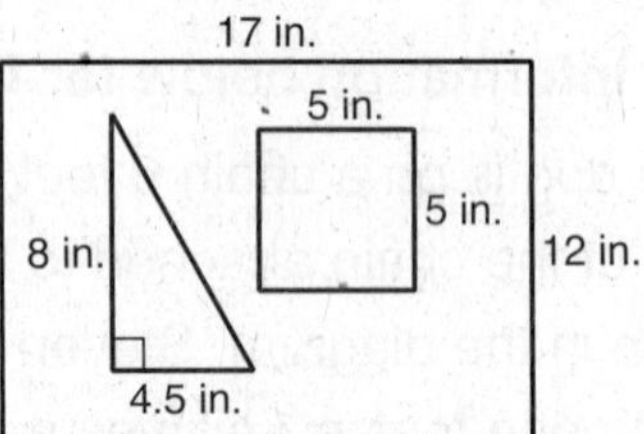

Choose the best answer.

5. A point is chosen randomly on $\overline{JM}$. Find the probability that the point is on $\overline{JK}$ or $\overline{JL}$. Round to the nearest hundredth.

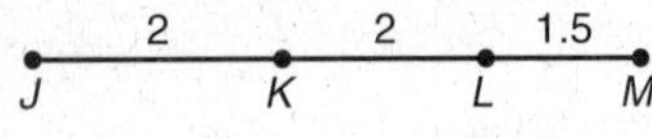

  A 0.41   **C** 0.81

  B 0.73   **D** 1.08

6. A train crosses at a railroad crossing 6 times a day—once every 4 hours. It takes an average of 5 minutes for the railroad gates to go down and then come up again. If you are approaching the railroad crossing, what is the probability that the gates are down?

  F $\dfrac{1}{360}$ **H** $\dfrac{1}{48}$

  G $\dfrac{1}{120}$ **J** $\dfrac{1}{30}$

7. On the dart board, the center circle has a diameter of 2 inches. What is the probability of hitting the shaded ring? Round to the nearest hundredth.

  A 0.01

  B 0.29

  C 0.30

  D 0.45

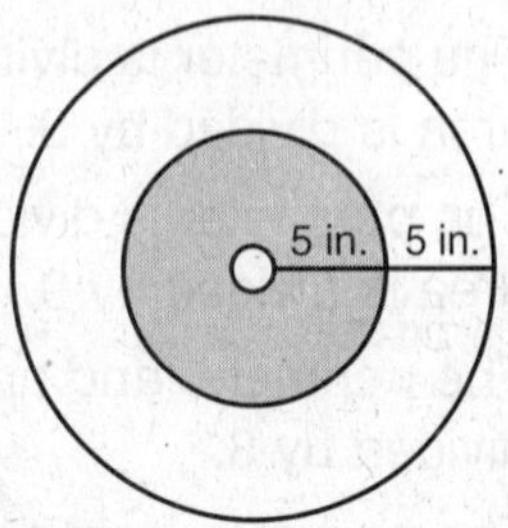

8. What is the probability that a coin randomly tossed into the rectangular fountain lands on one of the square "islands"? The "islands" are all the same size.

  F 0.03

  G 0.05

  H 0.15

  J 0.17

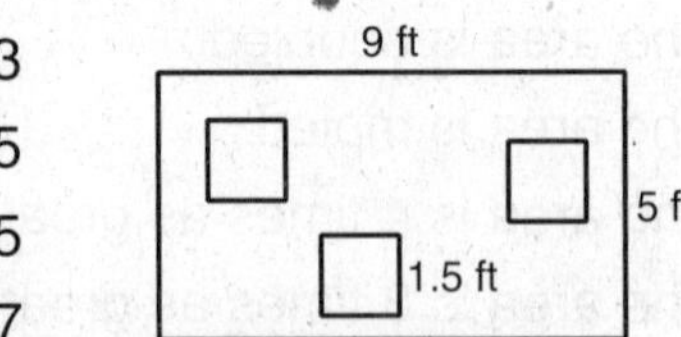

Holt Geometry

Problem Solving

LESSON 10-1

Solid Geometry

1. A slice of cheese is cut from the cylinder-shaped cheese as shown. Describe the cross section.

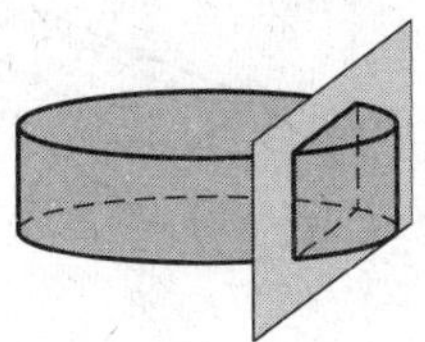

2. Mara has cut out five pieces of fabric to sew together to form a pillow. There are three rectangular pieces and two triangles. Describe the solid that will be formed.

____

3. A square pyramid is intersected by a plane as shown. Describe the cross section.

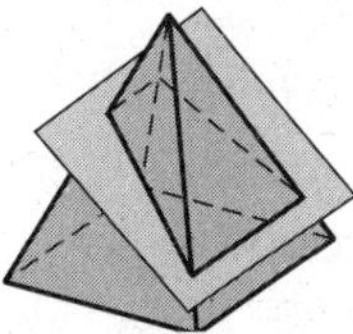

____

Choose the best answer.

4. A gift box is in the shape of a pentagonal prism. How many faces, edges, and vertices does the box have?

A 6 faces, 10 edges, 6 vertices

B 7 faces, 12 edges, 10 vertices

C 7 faces, 15 edges, 10 vertices

D 8 faces, 18 edges, 12 vertices

5. Which two solids have the same number of vertices?

F rectangular prism and triangular pyramid

G triangular prism and rectangular pyramid

H rectangular prism and pentagonal pyramid

J triangular prism and pentagonal pyramid

6. Which three-dimensional figure does the net represent?

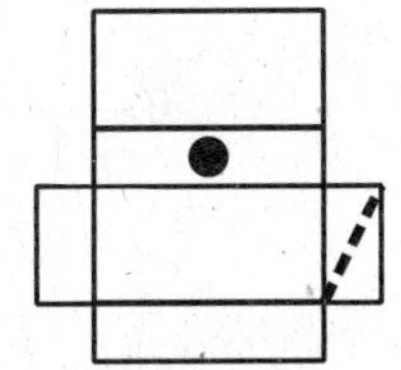

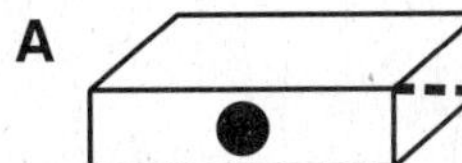
A

C

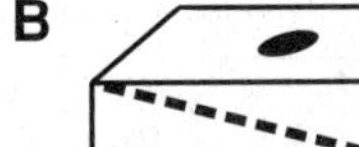
B

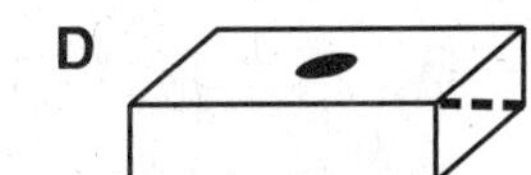
D

7. Which can be a true statement about the triangular prism whose net is shown?

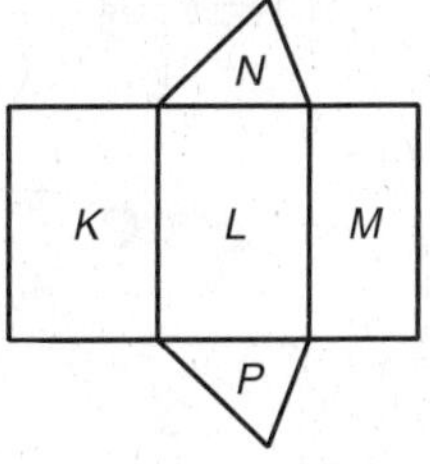

F Faces *L* and *M* are perpendicular.

G Faces *N* and *P* are perpendicular.

H Faces *K* and *L* are parallel.

J Faces *N* and *P* are parallel.

Holt Geometry

Problem Solving
Representations of Three-Dimensional Figures

1. Describe the top, front, and side views of the figure.

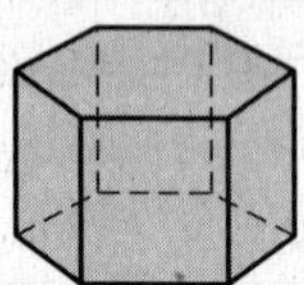

2. Erica used perspective to design the figure for a new logo. Describe the figure.

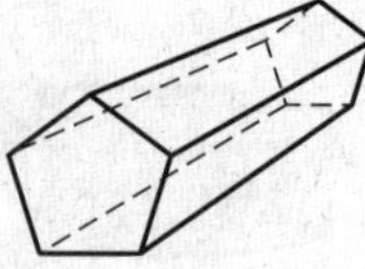

Choose the best answer.

3. Which is a true statement about the figure?

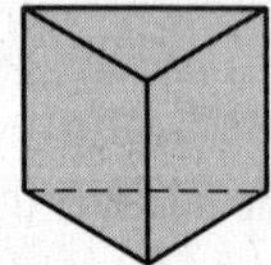

A The top view is a rectangle.
B A side view is a rectangle.
C A side view is a triangle.
D The front view is a triangle.

4. Which three-dimensional figure has these three views?

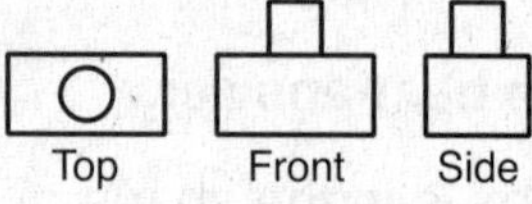

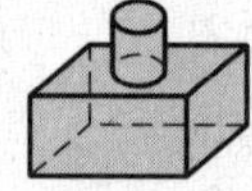
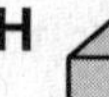
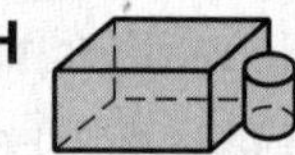

F 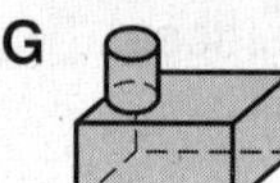   **H** 

G 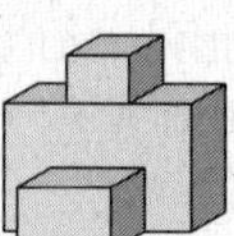   **J**

5. Which drawing best represents the top view of the three-dimensional figure? Assume there are no hidden cubes.

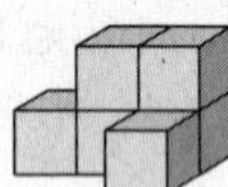

A  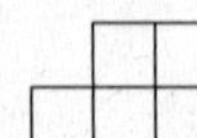   **C** 

B  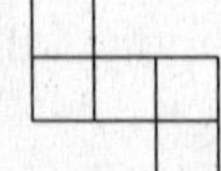   **D** 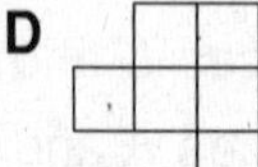

6. Which drawing best represents the side view of the building shown?

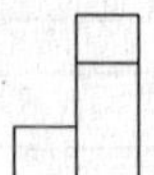

F   **H**

G 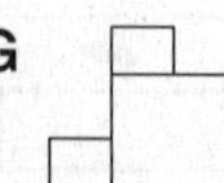   **J**

 # Problem Solving
Formulas in Three Dimensions

1. What is the height of the rectangular prism? Round to the nearest tenth if necessary.

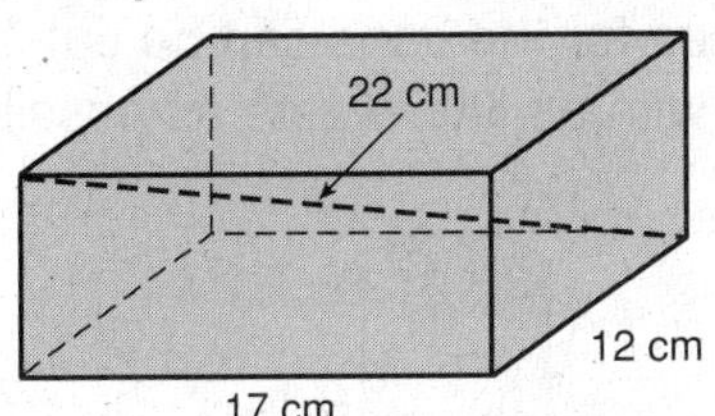

2. After lunch, Justin leaves the cafeteria to go to class, which is 22 feet north and 15 feet west of where he ate. The classroom is on the second floor, so it is 10 feet above the cafeteria. What is the actual distance between where Justin ate lunch and the classroom? Round to the nearest tenth.

3. Emily's hotel room is 18 feet south and 40 feet west of the pool. Her cousin Amber's hotel room is 22 feet north, 45 feet east, and 20 feet up on the third floor. How far apart are Emily's and Amber's rooms? Round to the nearest tenth.

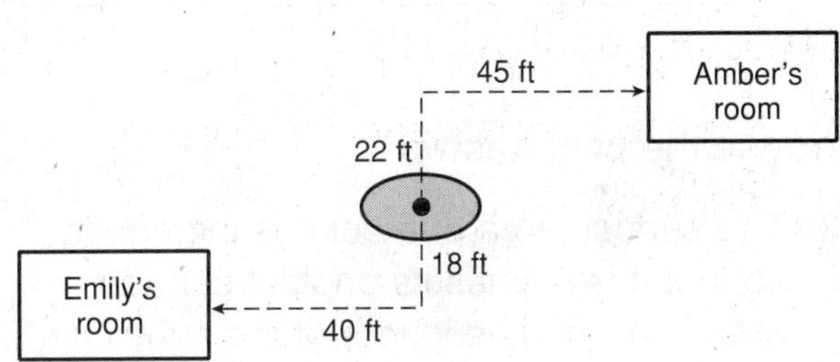

Choose the best answer.

4. How many faces, edges, and vertices does an octagonal pyramid have?

   A 7 faces, 12 edges, 7 vertices

   B 9 faces, 12 edges, 8 vertices

   C 9 faces, 16 edges, 9 vertices

   D 10 faces, 24 edges, 16 vertices

5. Which does NOT describe a polyhedron?

   F 8 vertices, 12 edges, 6 faces

   G 8 vertices, 10 edges, 6 faces

   H 6 vertices, 9 edges, 5 faces

   J 6 vertices, 10 edges, 6 faces

6. Point R has coordinates (8, 6, 1), and the midpoint of $\overline{RS}$ is M(15, −2, 7). Which is the best estimate for the distance between point R and point S?

   A 10.0 units   C 21.0 units

   B 12.2 units   D 24.4 units

7. A rectangular prism has the following vertices. What is the volume of the prism?

   A(0, 0, 4) B(−4, 0, 0)

   C(−4, 2, 0) D(0, 2, 0)

   E(0, 0, 0) F(−4, 0, 4)

   G(−4, 2, 4) H(0, 2, 4)

   F 4 units^3 H 32 units^3

   G 16 units^3 J 64 units^3

Holt Geometry

Problem Solving
Surface Area of Prisms and Cylinders

1. The lateral area of the regular pentagonal prism below is 220 mm^2. What is the surface area? Round to the nearest tenth if necessary.

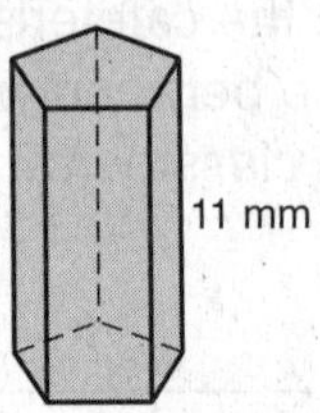

____

2. A sheet of metal 8 feet long and 6 feet wide is to be cut into cylindrical cans like the one shown. How many lateral surfaces for the cans can be cut from the metal with as little waste as possible?

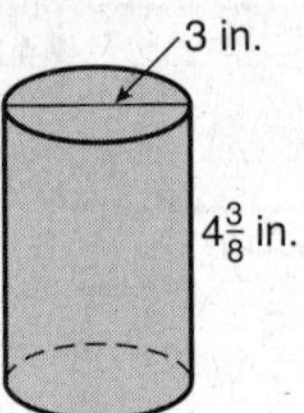

____

Choose the best answer.

3. The surface area of a cube is increased so that it is 9 times its original surface area. How did the length of the cube change?

 A The length was doubled.

 B The length was tripled.

 C The length was quadrupled.

 D The length was multiplied by 9.

4. A rectangular prism has a surface area of 152 square inches. If the length, width, and height are all changed to $\frac{1}{2}$ their original size, what will be the new surface area of the prism?

 F 19 in^2

 G 38 in^2

 H 76 in^2

 J 114 in^2

5. Determine the surface area exposed to the air of the composite figure shown. Round to the nearest tenth.

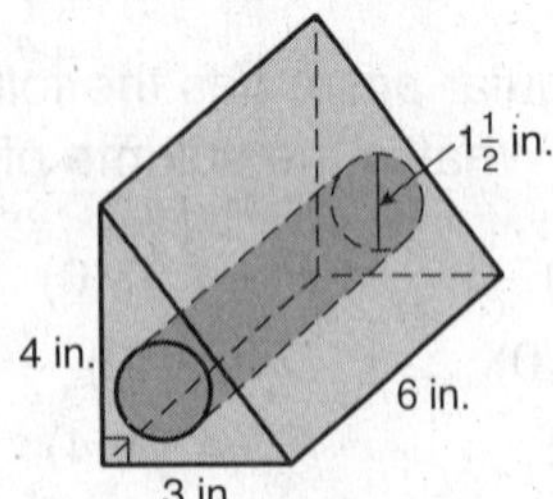

 A 98.1 in^2

 B 107.6 in^2

 C 108.7 in^2

 D 110.5 in^2

6. Which of the two cylindrical cans has a greater surface area?

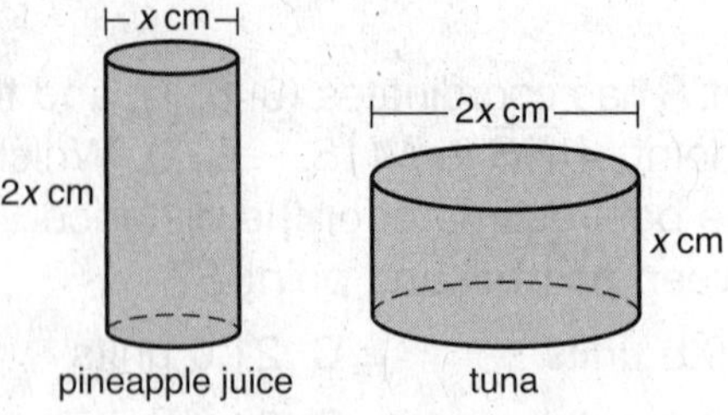

 F pineapple juice can

 G tuna can

 H The two cans have the same surface area.

 J It is impossible to determine which can has a greater surface area.

Holt Geometry

Problem Solving
Surface Area of Pyramids and Cones

1. Find the diameter of a right cone with slant height 18 centimeters and surface area 208π square centimeters.

2. Find the surface area of a regular pentagonal pyramid with base area 49 square meters and slant height 13 meters. Round to the nearest tenth.

3. A piece of paper in the shape shown is folded to form a cone. What is the diameter of the base of the cone that is formed? Round to the nearest tenth.

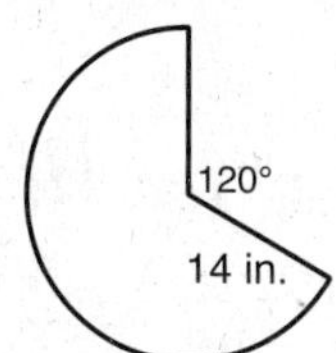

4. The right cone has a surface area of 240π square millimeters. What is the radius of the cone?

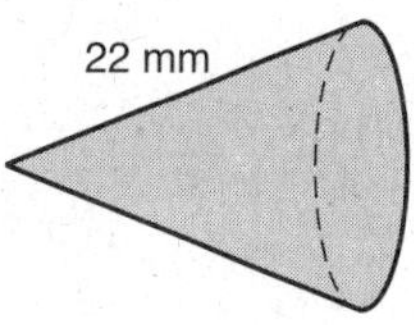

Choose the best answer.

5. A square pyramid has a base with a side length of 9 centimeters and a slant height that is 4 centimeters more than $1\frac{1}{2}$ times the length of the base. Find the surface area of the pyramid.

   A 162 cm²   **C** 315 cm²
   B 243 cm²   **D** 396 cm²

6. A cone has a surface area of 64π square inches. If the radius and height are each multiplied by $\frac{3}{4}$, what will be the new surface area of the cone?

   F 36π in²   **H** 60π in²
   G 48π in²   **J** 96π in²

7. Find the surface area of the composite figure. Round to the nearest tenth.

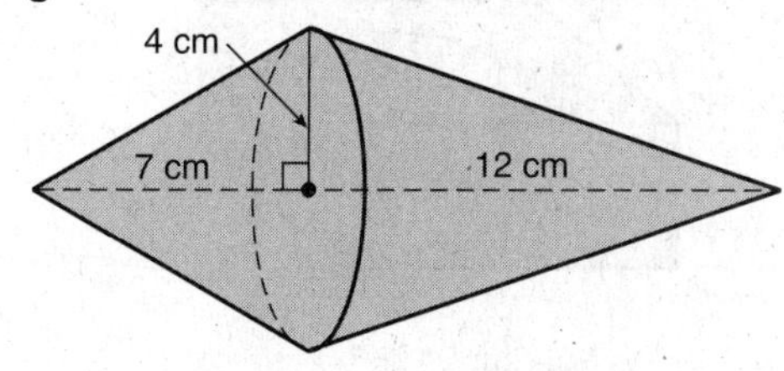

   A 238.8 cm²   **C** 311.0 cm²
   B 260.3 cm²   **D** 361.3 cm²

8. A cone has a base diameter of 6 yards. What is the slant height of the cone if it has the same surface area as the square pyramid shown? Round to the nearest tenth.

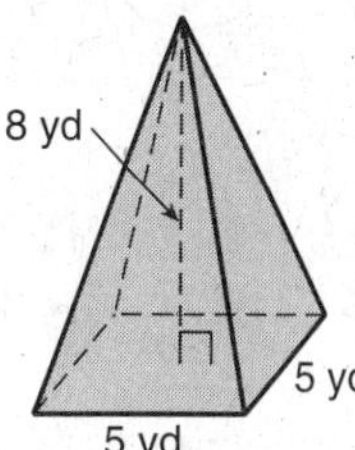

   F 8.1 yd   **H** 11.3 yd
   G 8.5 yd   **J** 25.6 yd

Holt Geometry

Problem Solving
Volume of Prisms and Cylinders

1. A cylindrical juice container has the dimensions shown. About how many cups of juice does this container hold? (*Hint:* 1 cup ≈ 14.44 in^3)

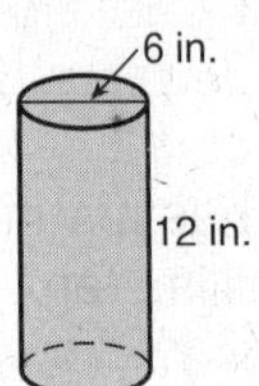

2. A large cylindrical cooler is $2\frac{1}{2}$ feet high and has a diameter of $1\frac{1}{2}$ feet. It is filled $\frac{3}{4}$ high with water for athletes to use during their soccer game. Estimate the volume of the water in the cooler in gallons. (*Hint:* 1 gallon ≈ 231 in^3)

____

Choose the best answer.

3. How many 3-inch cubes can be placed inside the box?

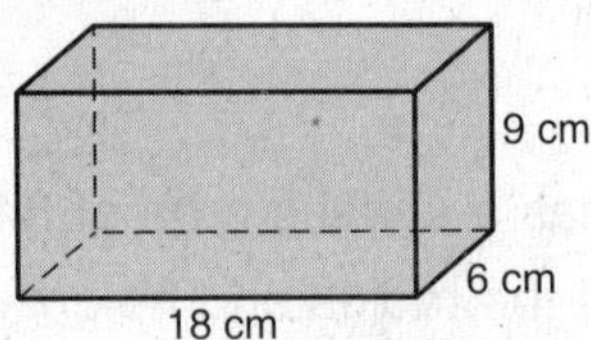

A 27   **C** 45

B 36   **D** 72

4. A cylinder has a volume of 4π cm^3. If the radius and height are each tripled, what will be the new volume of the cylinder?

F 12π cm^3 **H** 64π cm^3

G 36π cm^3 **J** 108π cm^3

5. What is the volume of the composite figure with the dimensions shown in the three views? Round to the nearest tenth.

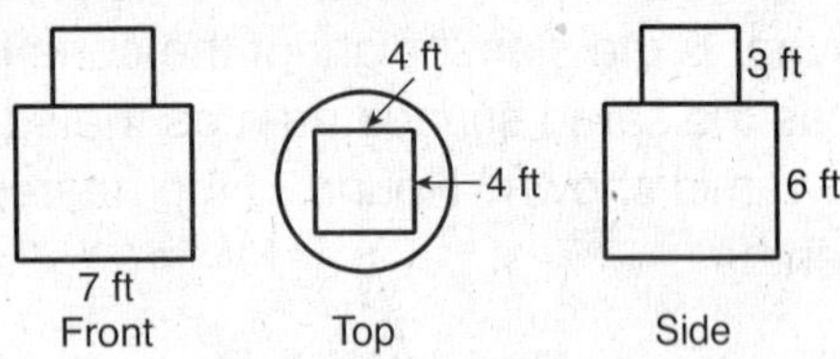

A 182.9 ft^3 **C** 278.9 ft^3

B 205.7 ft^3 **D** 971.6 ft^3

6. Find the expression that can be used to determine the volume of the composite figure shown.

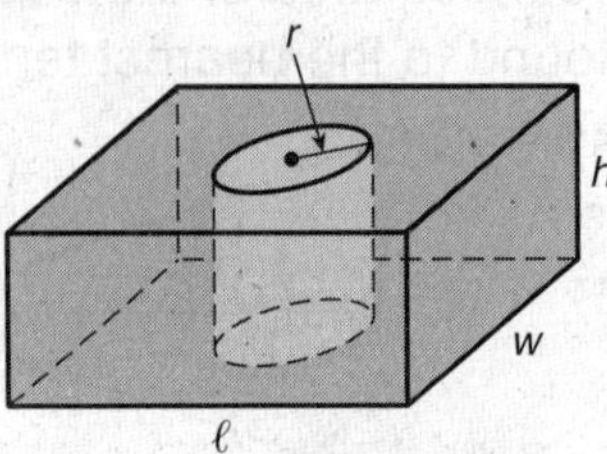

F $\ell wh - \pi r^2 h$ **H** $\pi r^2 h - \ell wh$

G $\pi r^2 h + \ell wh$ **J** $\ell wh + 2\pi r^2 h$

Holt Geometry

Problem Solving
Volume of Pyramids and Cones

1. A regular square pyramid has a base area of 196 meters and a lateral area of 448 square meters. What is the volume of the pyramid? Round your answer to the nearest tenth.

____

2. A paper cone for serving roasted almonds has a volume of 406π cubic centimeters. A smaller cone has half the radius and half the height of the first cone. What is the volume of the smaller cone? Give your answer in terms of π.

____

3. The hexagonal base in the pyramid is a regular polygon. What is the volume of the pyramid if its height is 9 centimeters? Round to the nearest tenth.

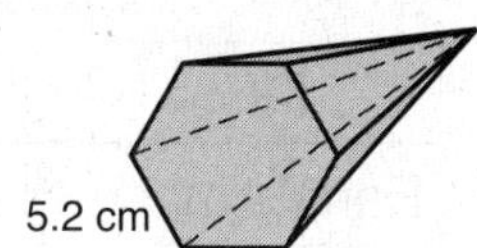

____

4. Find the volume of the shaded solid in the figure shown. Give your answer in terms of π.

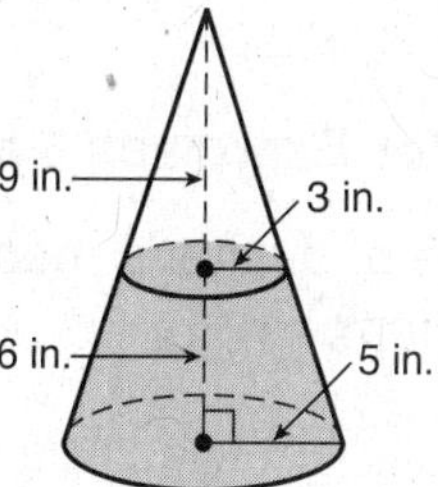

____

Choose the best answer.

5. The diameter of the cone equals the width of the cube, and the figures have the same height. Find the expression that can be used to determine the volume of the composite figure.

   A $4(4)(4) - \frac{1}{3}\pi(2^2)(4)$

   B $4(4)(4) + \frac{1}{3}\pi(2^2)(4)$

   C $4(4)(4) - \pi(2^2)(4)$

   D $4(4)(4) + \frac{1}{3}\pi(2^2)$

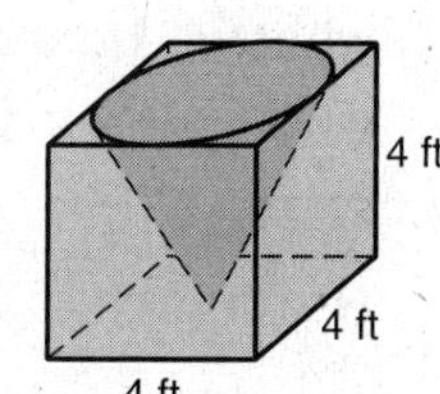

7. The Step Pyramid of Djoser in Lower Egypt was the first pyramid in the history of architecture. Its original height was 204 feet, and it had a rectangular base measuring 411 feet by 358 feet. Which is the best estimate for the volume of the pyramid in cubic yards?

   A $370{,}570 \text{ yd}^3$

   C $3{,}335{,}128 \text{ yd}^3$

   B $1{,}111{,}709 \text{ yd}^3$

   D $10{,}005{,}384 \text{ yd}^3$

6. Approximately how many fluid ounces of water can the paper cup hold? (*Hint:* 1 fl oz $\approx 1.805 \text{ in}^3$)

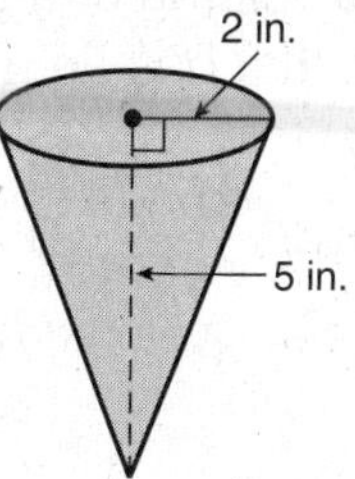

   F 10.9 fl oz

   H 32.7 fl oz

   G 11.6 fl oz

   J 36.3 fl oz

Holt Geometry

LESSON 10-8 Problem Solving
Spheres

1. A globe has a volume of 288π in^3. What is the surface area of the globe? Give your answer in terms of π.

____

2. Eight bocce balls are in a box 18 inches long, 9 inches wide, and 4.5 inches deep. If each ball has a diameter of 4.5 inches, what is the volume of the space around the balls? Round to the nearest tenth.

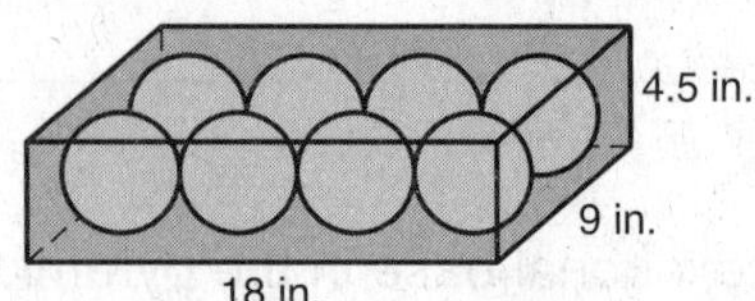

Use the table for Exercises 3 and 4.

Ganymede, one of Jupiter's moons, is the largest moon in the solar system.

Moon	Diameter
Earth's moon	2160 mi
Ganymede	3280 mi

3. Approximately how many times as great as the volume of Earth's moon is the volume of Ganymede?

____

4. Approximately how many times as great is the surface area of Ganymede than the surface area of Earth's moon?

____

Choose the best answer.

5. What is the volume of a sphere with a great circle that has an area of 225π cm^2?

  A 300π cm^3 **C** 2500π cm^3

  B 900π cm^3 **D** 4500π cm^3

6. A hemisphere has a surface area of 972π cm^2. If the radius is multiplied by $\frac{1}{3}$, what will be the surface area of the new hemisphere?

  F 36π cm^2 **H** 162π cm^2

  G 108π cm^2 **J** 324π cm^2

7. Which expression represents the volume of the composite figure formed by the hemisphere and cone?

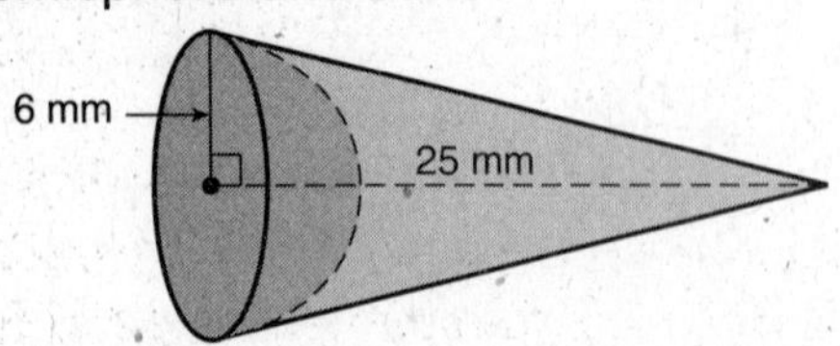

  A 52π mm^3 **C** 276π mm^3

  B 156π mm^3 **D** 288π mm^3

8. Which best represents the surface area of the composite figure?

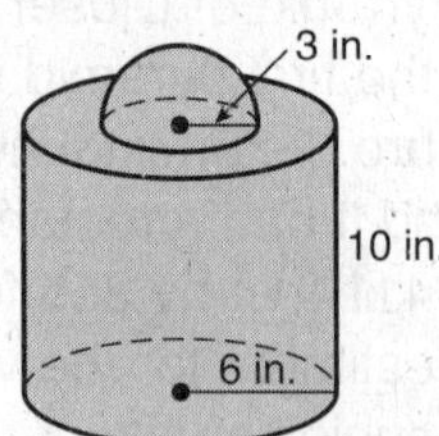

  F 129π in^2 **H** 201π in^2

  G 138π in^2 **J** 210π in^2

Holt Geometry

LESSON 11-1 Problem Solving
Lines That Intersect Circles

1. The cruising altitude of a commercial airplane is about 9000 meters. Use the diagram to find *AB*, the distance from an airplane at cruising altitude to Earth's horizon. Round to the nearest kilometer.

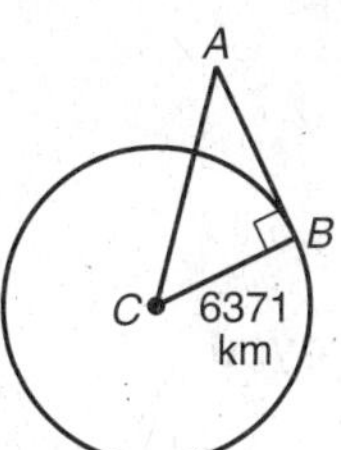

2. In the figure, segments that appear to be tangent are tangent. Find *QS*.

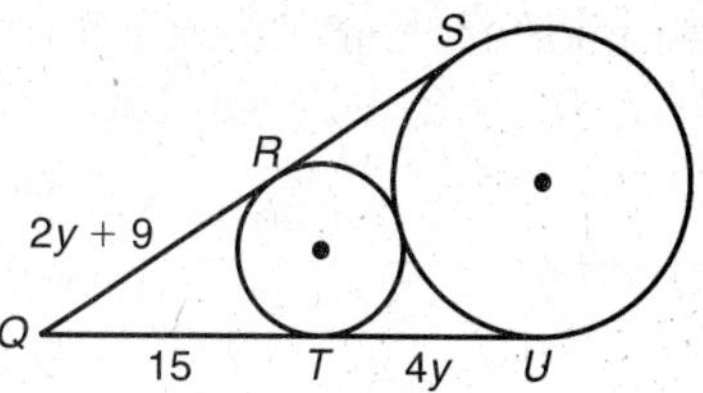

3. The area of $\odot H$ is 100π, and $HF = 26$ centimeters. What is the perimeter of quadrilateral *EFGH*?

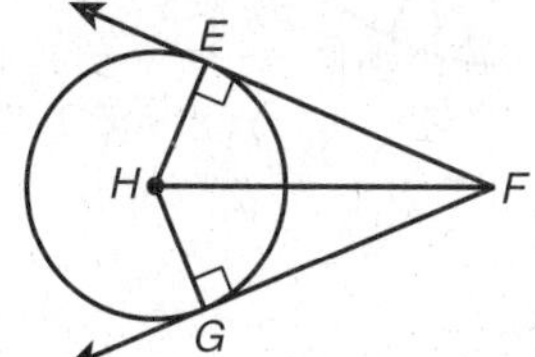

4. $\overrightarrow{IH}$, $\overrightarrow{IK}$, and $\overrightarrow{KL}$ are tangent to $\odot A$. What is *IK*?

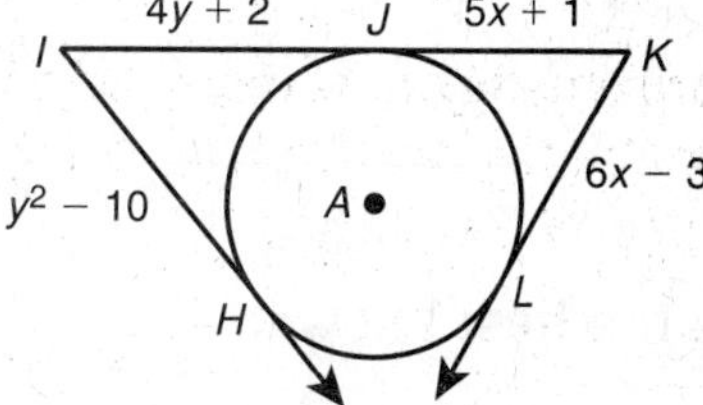

Choose the best answer.

5. A teardrop-shaped roller coaster loop is a section of a spiral in which the radius is constantly changing. The radius at the bottom of the loop is much larger than the radius at the top of the loop, as shown in the figure. Which is a true statement?

A $\odot K$ and $\odot M$ have two points of tangency.

B $\odot K$, $\odot L$, and $\odot M$ have one point of tangency.

C $\odot L$ is internally tangent to $\odot K$ and $\odot M$.

D $\odot L$ is externally tangent to $\odot K$ and $\odot M$.

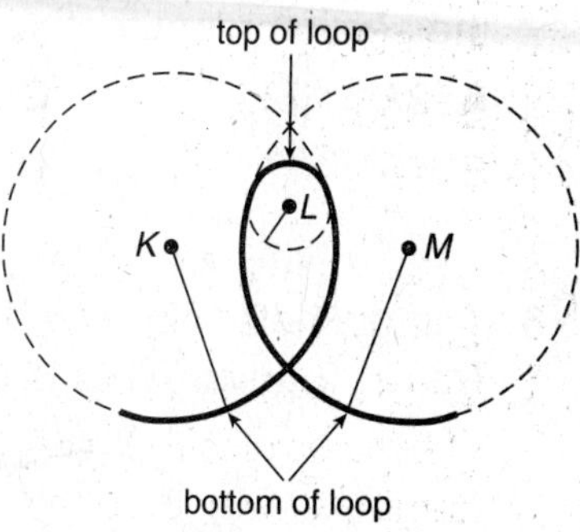

6. $\odot G$ has center (2, 5) and radius 3. $\odot H$ has center (2, 0). If the circles are tangent, which line could be tangent to both circles?

F $x = 2$

G $x = 0$

H $y = 2$

J $y = 5$

7. The Hubble Space Telescope orbits 353 miles above Earth, and Earth's radius is about 3960 miles. Which is closest to the distance from the telescope to Earth's horizon?

A 1634 mi

B 1709 mi

C 3976 mi

D 5855 mi

Holt Geometry

Name _________________________________ Date __________ Class __________

Problem Solving
Arcs and Chords

1. Circle D has center $(-2, -7)$ and radius 7. What is the measure, in degrees, of the major arc that passes through points $H(-2, 0)$, $J(5, -7)$, and $K(-9, -7)$?

2. A circle graph is composed of sectors with central angles that measure $3x°$, $3x°$, $4x°$, and $5x°$. What is the measure, in degrees, of the smallest minor arcs?

Use the following information for Exercises 3 and 4.

The circle graph shows the results of a survey in which teens were asked what says the most about them at school. Find each of the following.

3. $\text{m}\overset{\frown}{AB}$

____

4. $\text{m}\angle APC$

____

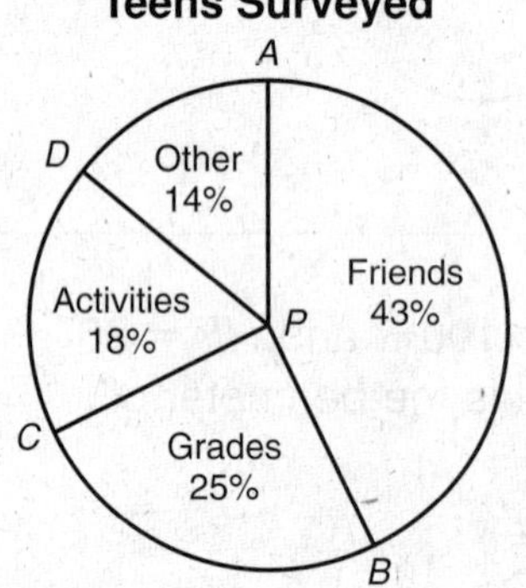

Choose the best answer.

5. Students were asked to name their favorite cafeteria food. The results of the survey are shown in the table. In a circle graph showing these results, which is closest to the measure of the central angle for the section representing chicken tenders?

   A 21°   **C** 83°

   B 75°   **D** 270°

Favorite Lunch	Number of Students
Pizza	108
Chicken tenders	75
Taco salad	90
Other	54

6. The diameter of $\odot R$ is 15 units, and $HJ = 12$ units. What is the length of $\overline{ST}$?

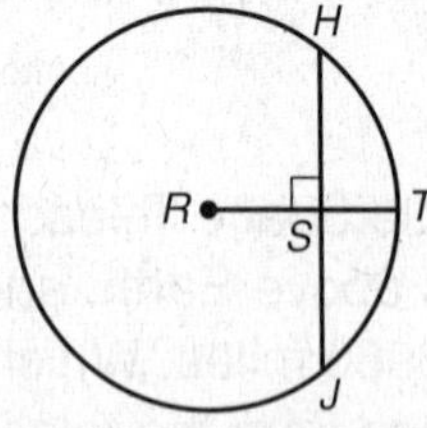

   F 2.1 units   **H** 4.5 units

   G 3 units   **J** 9.6 units

7. In the stained glass window, $\overline{AB} \cong \overline{CD}$ and $\overline{AB} \parallel \overline{CD}$. What is $\text{m}\overset{\frown}{CBD}$?

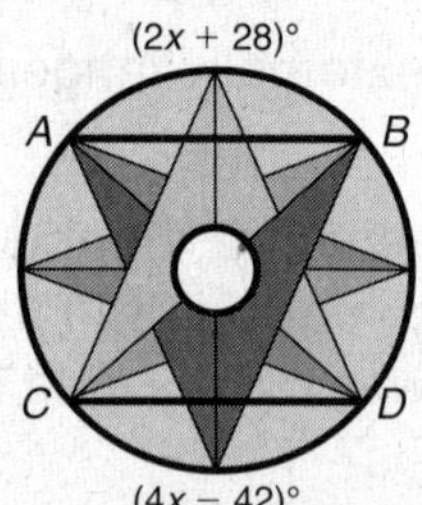

   A 35°   **C** 98°

   B 70°   **D** 262°

Holt Geometry

Problem Solving
Sector Area and Arc Length

1. A circle with a radius of 20 centimeters has a sector that has an arc measure of 105°. What is the area of the sector? Round to the nearest tenth.

____

2. A sector whose central angle measures 72° has an area of 16.2π square feet. What is the radius of the circle?

____

3. The archway below is to be painted. What is the area of the archway to the nearest tenth?

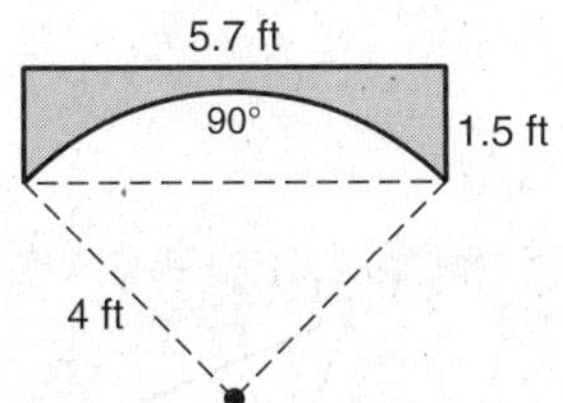

4. Circle *N* has a circumference of 16π millimeters. What is the area of the shaded region to the nearest tenth?

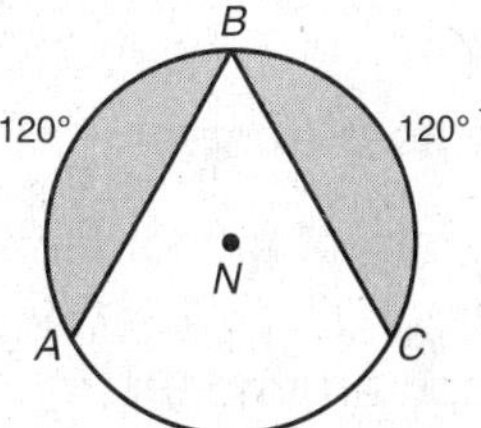

Choose the best answer.

5. The circular shelves in diagram are each 28 inches in diameter. The "cut-out" portion of each shelf is 90°. Approximately how much shelf paper is needed to cover both shelves?

  A 154 in^2

  B 308 in^2

  C 462 in^2

  D 924 in^2

6. Find the area of the shaded region. Round to the nearest tenth.

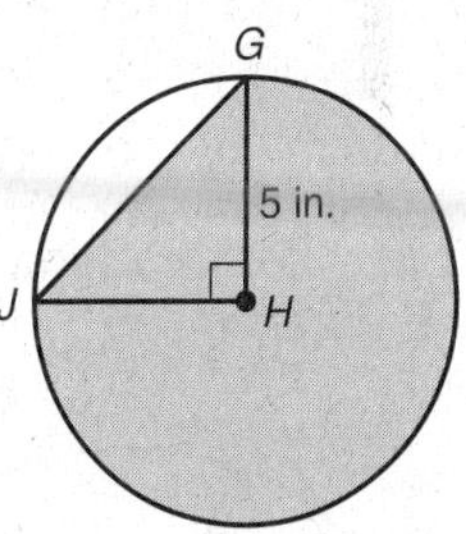

  F 8.2 in^2 **H** 71.4 in^2

  G 19.6 in^2 **J** 78.5 in^2

7. A semicircular garden with a diameter of 6 feet is to have 2 inches of mulch spread over it. To the nearest tenth, what is the volume of mulch that is needed?

  A 2.4 ft^3 **C** 14.1 ft^3

  B 4.8 ft^3 **D** 28.3 ft^3

8. A round cheesecake 12 inches in diameter and 3 inches high is cut into 8 equal-sized pieces. If five pieces have been taken, what is the approximate volume of the cheesecake that remains?

  F 42.4 in^3 **H** 127.2 in^3

  G 70.7 in^3 **J** 212.1 in^3

Holt Geometry

<table><tr><td>**LESSON**
11-4</td><td># Problem Solving
Inscribed Angles</td></tr></table>

1. Find m$\widehat{AB}$.

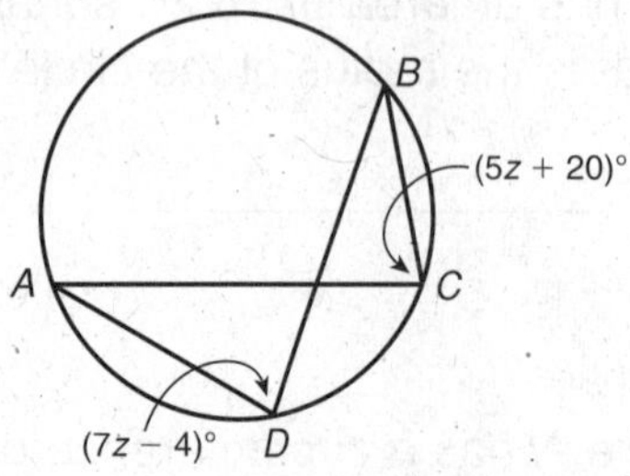

2. Find the angle measures of *RSTU*.

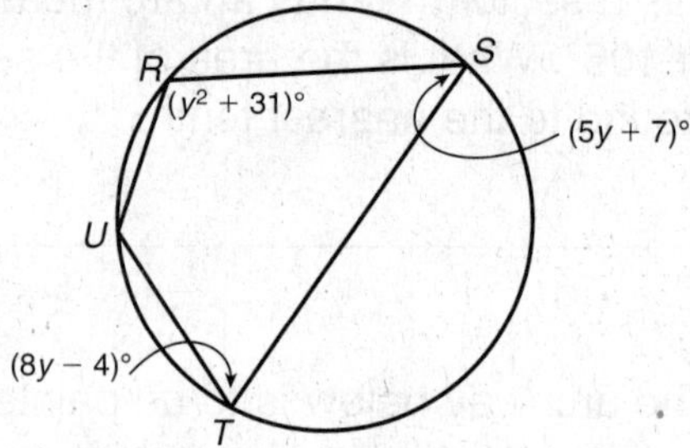

Choose the best answer.

Use the diagram of a floor tile for Exercises 3 and 4. Points *Q, R, S, T, U, V, W,* and *X* are equally spaced around ⊙*L*.

3. Find m∠*RQT*.

  A 15°   **C** 45°

  B 30°   **D** 60°

4. Find m∠*QRS*.

  F 67.5°   **H** 180°

  G 135°   **J** 270°

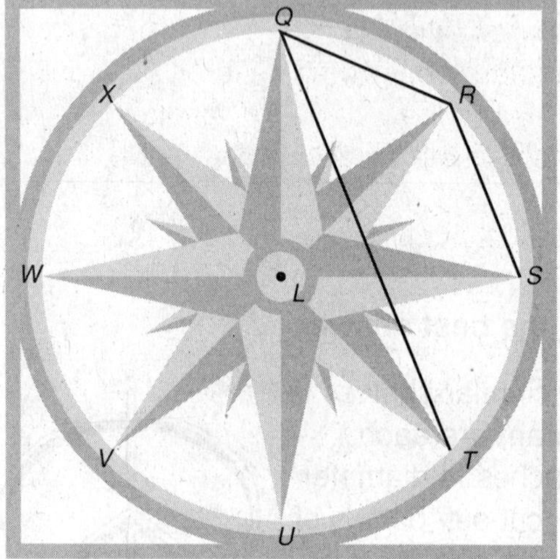

5. If m∠*KLM* = 20° and m$\widehat{MP}$ = 30°, what is m∠*KNP*?

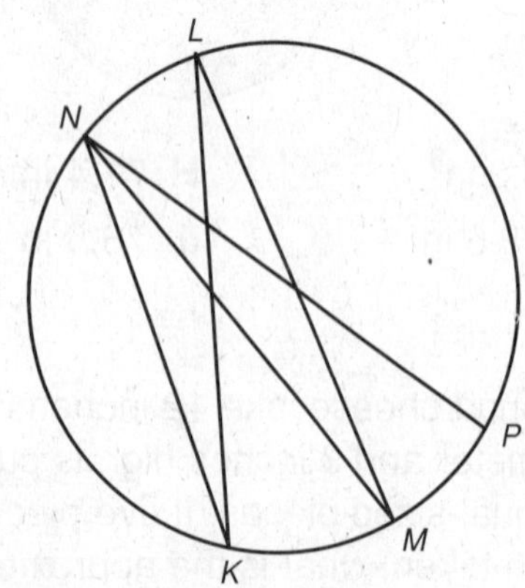

  A 25°   **C** 50°

  B 35°   **D** 70°

6. In ⊙*M*, m∠*AMB* = 74°. What is m∠*CDB*?

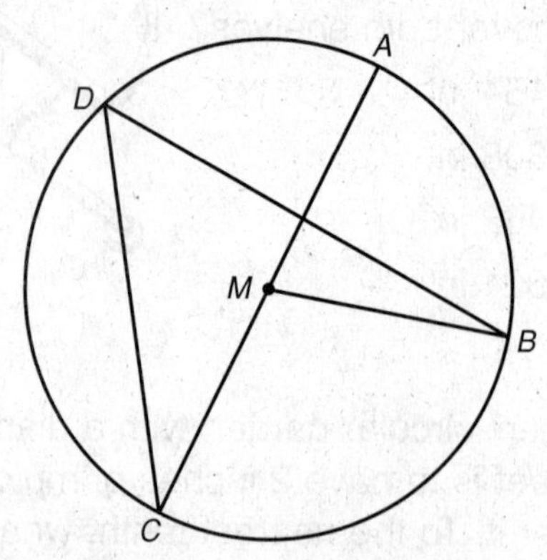

  F 37°   **H** 74°

  G 53°   **J** 106°

Holt Geometry

LESSON 11-5 — Problem Solving
Angle Relationships in Circles

1. What is $m\widehat{LM}$?

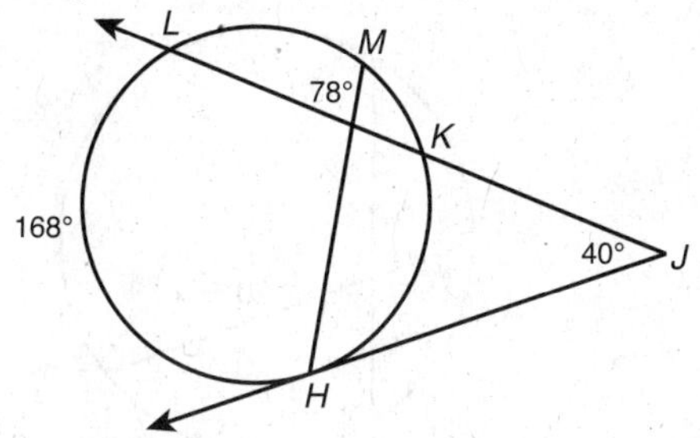

2. An artist painted the design shown below. What is the value of x?

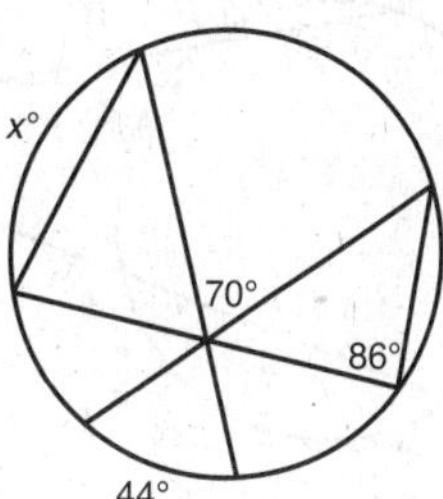

For Exercises 3 and 4, use the diagrams.

3. A polar orbiting satellite is about 850 kilometers above Earth. About 69.2 arc degrees of the planet are visible to a camera in the satellite. What is $m\angle P$?

____

4. A geostationary satellite is about 35,800 kilometers above Earth. How many arc degrees of the planet are visible to a camera in the satellite?

____

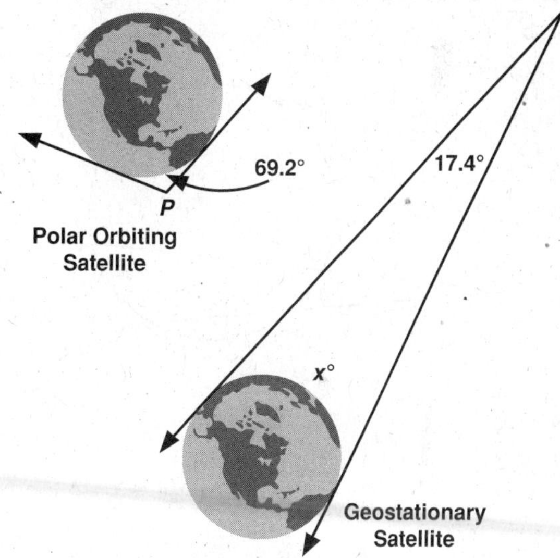

Choose the best answer.

5. What is $m\angle ADE$?

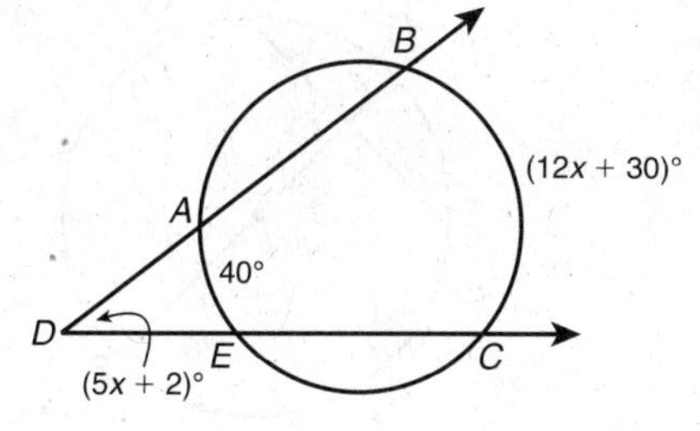

A 7°   **C** 37°

B 33°   **D** 114°

6. Find $m\angle VTU$.

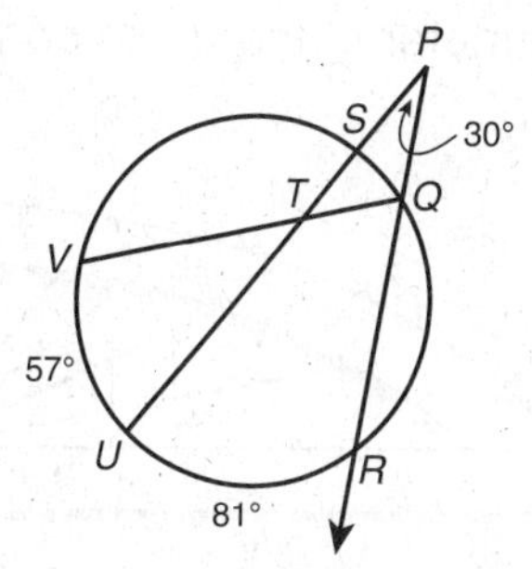

F 21°   **H** 36°

G 29°   **J** 39°

Holt Geometry

Problem Solving
Segment Relationships in Circles

1. Find *EG* to the nearest tenth.

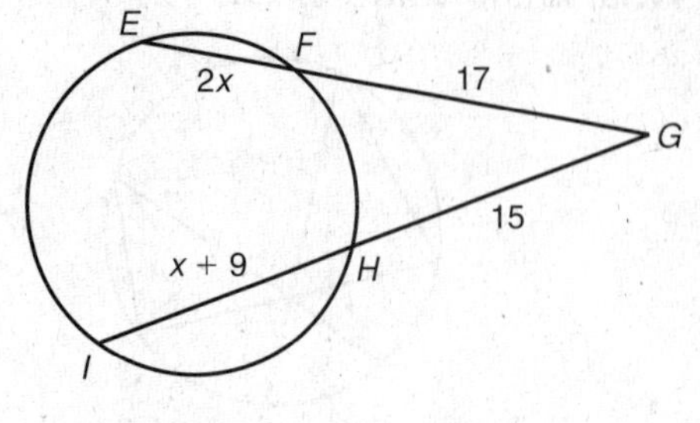

2. What is the length of $\overline{UW}$?

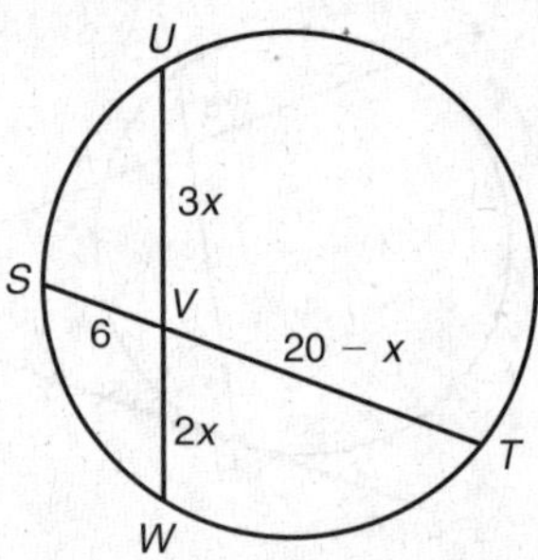

____

____

Choose the best answer.

3. Which of these is closest to the length of $\overline{ST}$?

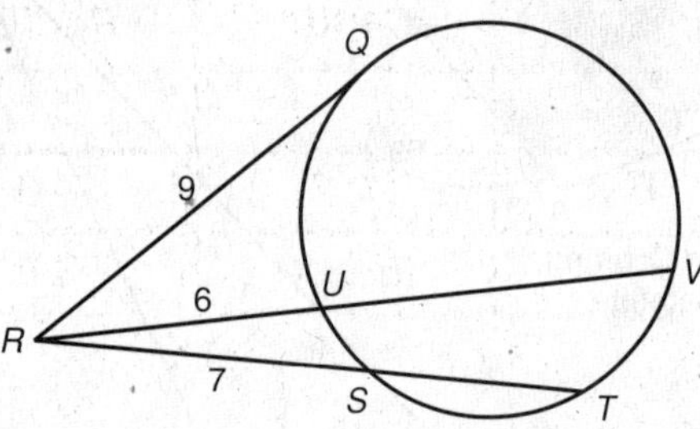

A 4.6   **C** 7.5

B 5.4   **D** 11.6

4. Floral archways like the one shown below are going to be used for the prom. $\overline{LN}$ is the perpendicular bisector of $\overline{KM}$. $KM = 6$ feet and $LN = 2$ feet. What is the diameter of the circle that contains $\overset{\frown}{KM}$?

F 4.5 ft

G 5.5 ft

H 6.5 ft

J 8 ft

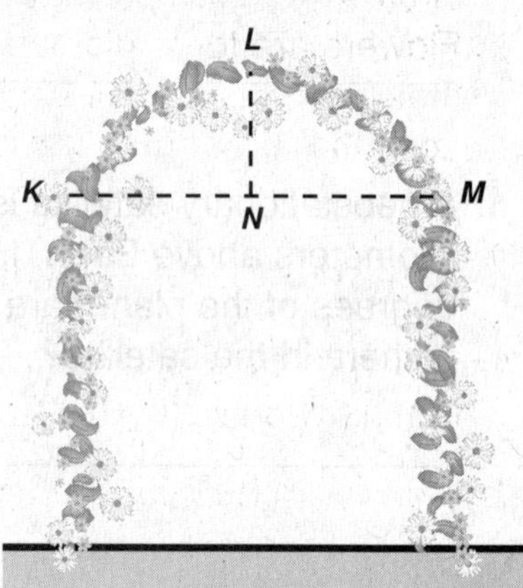

5. The figure is a "quarter" wood arch used in architecture. $\overline{WX}$ is the perpendicular bisector of the chord containing $\overline{YX}$. Find the diameter of the circle containing the arc.

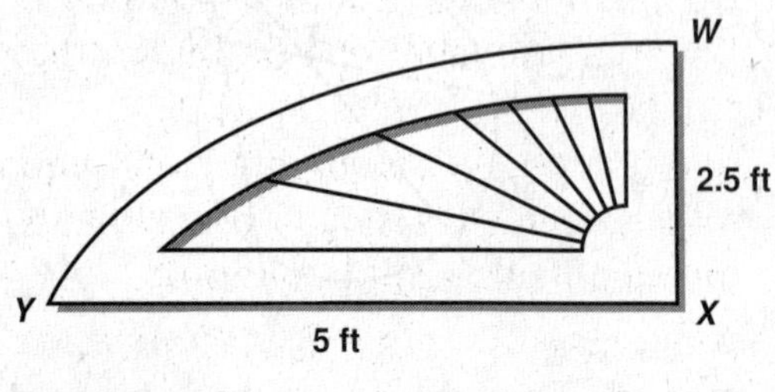

A 5 ft   **C** 10 ft

B 8.5 ft   **D** 12.5 ft

6. In $\odot N$, $CD = 18$. Find the radius of the circle to the nearest tenth.

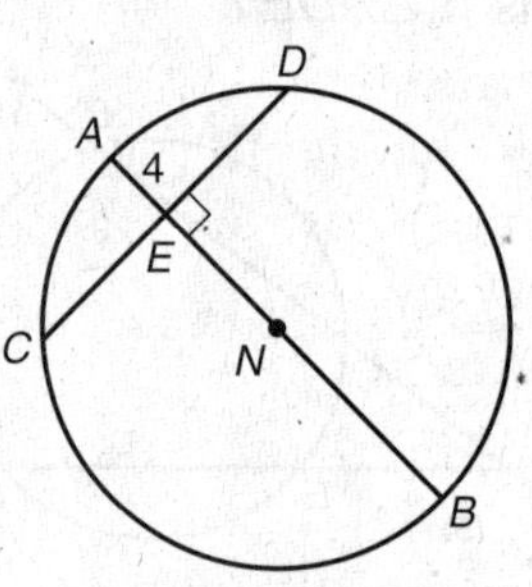

F 12.1   **H** 20.3

G 16.3   **J** 24.3

Holt Geometry

Problem Solving
Circles in the Coordinate Plane

1. Write the equation of the circle that contains the points graphed below.

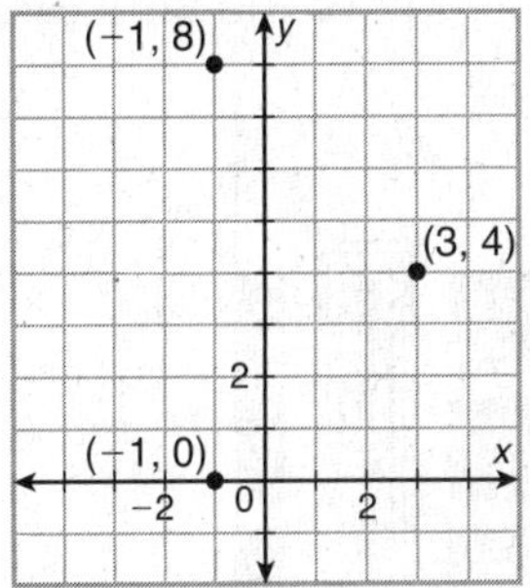

2. Find the area of a circle that has center *J* and passes through *K*. Express your answer in terms of π.

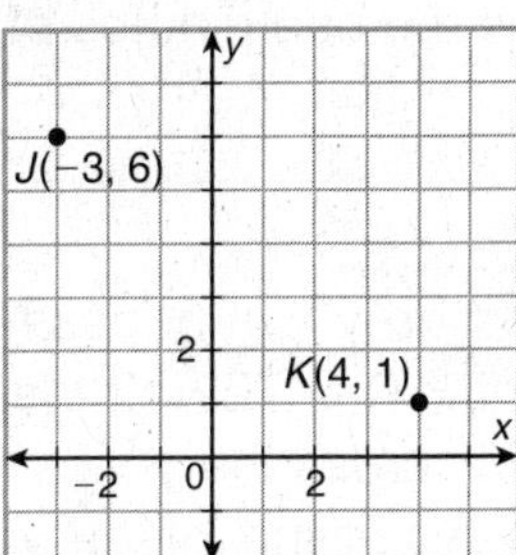

____

Choose the best answer.

3. An English knot garden has hedges planted to form geometric shapes. A blueprint of a knot garden contains three circular hedges as described in the table. Flowers are to be planted in the space that is within all three circles. Which is a point that could be planted with flowers?

Circular Hedge	Center	Radius
A	(3, 2)	3 ft
B	(7, 2)	4 ft
C	(5, −1)	3 ft

A (7, 1) **C** (0, 5)

B (5, 1) **D** (0, 0)

4. Which of these circles intersects the circle that has center (0, 6) and radius 1?

F $(x - 5)^2 + (y + 3)^2 = 4$

G $(x - 4)^2 + (y - 3)^2 = 9$

H $(x + 5)^2 + (y + 1)^2 = 16$

J $(x + 1)^2 + (y - 4)^2 = 4$

5. The center of $\odot S$ is (9, 2), and the radius of the circle is 5 units. Which is a point on the circle?

A (4, 2) **C** (9, 4)

B (14, 0) **D** (9, −5)

6. Which is an equation for a circle that has the same center as $\odot P$ but has a circumference that is four times as great?

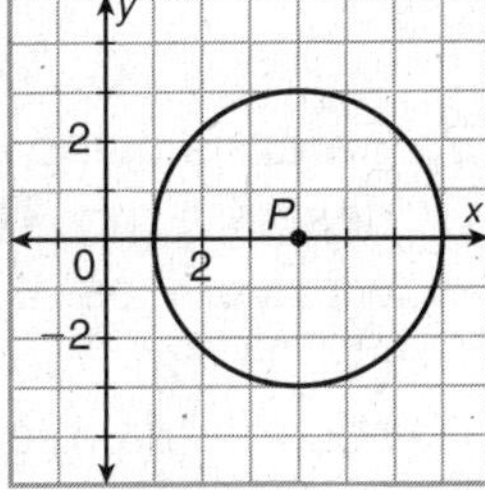

F $(x - 4)^2 + y^2 = 36$

G $(x - 4)^2 + y^2 = 144$

H $x^2 + (y - 4)^2 = 36$

J $x^2 + (y - 4)^2 = 144$

7. The Maxair amusement park ride consists of a circular ring that holds 50 riders. Suppose that the center of the ride is at the origin and that one of the riders on the circular ring is at (16, 15.1). If one unit on the coordinate plane equals 1 foot, which is a close approximation of the circumference of the ride?

A 22 ft **C** 138 ft

B 44 ft **D** 1521 ft

Holt Geometry

Problem Solving
Reflections

1. Quadrilateral *JLKM* has vertices *J*(7, 9), *K*(0, −4), *L*(2, 2), and *M*(5, −3). If the figure is reflected across the line $y = x$, what are the coordinates of *M′*?

____

3. The function $y = -3^x$ passes through the point *P*(6, −729). If the graph is reflected across the *y*-axis, what are the coordinates of the image of *P*?

____

2. In the drawing, the left side of a structure is shown with its line of reflection. Draw the right side of the structure.

Choose the best answer.

4. A park planner is designing two paths that connect picnic areas *E* and *F* to a point on the park road. Which point on the park road will make the total length of the paths as small as possible? (*Hint:* Use a reflection. What is the shortest distance between two points?)

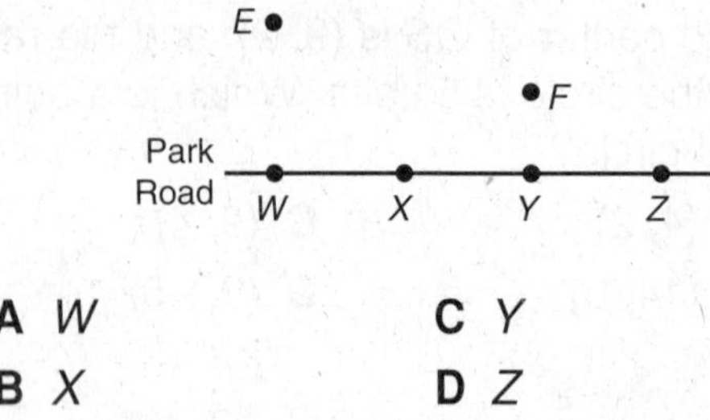

A *W*
B *X*
C *Y*
D *Z*

5. △*RST* is reflected across a line so that *T′* has coordinates (1, 3). What are the coordinates of *S′*?

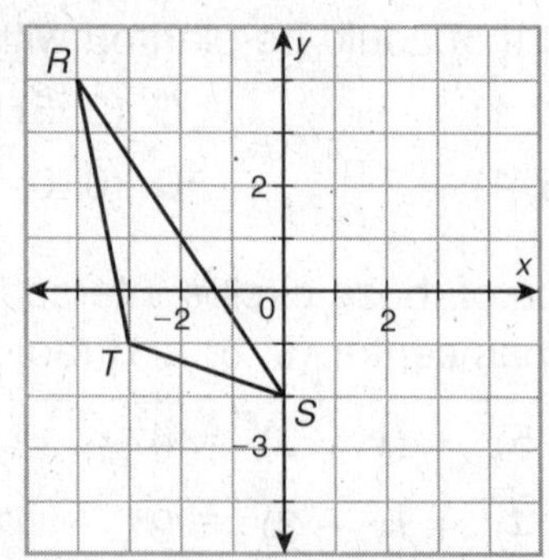

F (0, 2)
G (0, −2)
H (2, 0)
J (−2, 0)

6. △*MNP* with vertices *M*(1, 5), *N*(0, −3), and *P*(−2, 2) is reflected across a line. The coordinates of the reflection image are *M′*(7, 5), *N′*(8, −3), and *P′*(10, 2). Over which line was △*MNP* reflected?

A $y = 2$
B $x = 2$
C $y = 4$
D $x = 4$

7. Sarah is using a coordinate plane to design a rug. The rug is to have a triangle with vertices at (8, 13), (2, −13), and (14, −13). She wants the rug to have a second triangle that is the reflection of the first triangle across the *x*-axis. Which is a vertex of the second triangle?

F (−13, 14)
G (−14, 13)
H (−2, −13)
J (2, 13)

Holt Geometry

Problem Solving
Translations

1. A checker player's piece begins at K and, through a series of moves, lands on L. What translation vector represents the path from K to L?

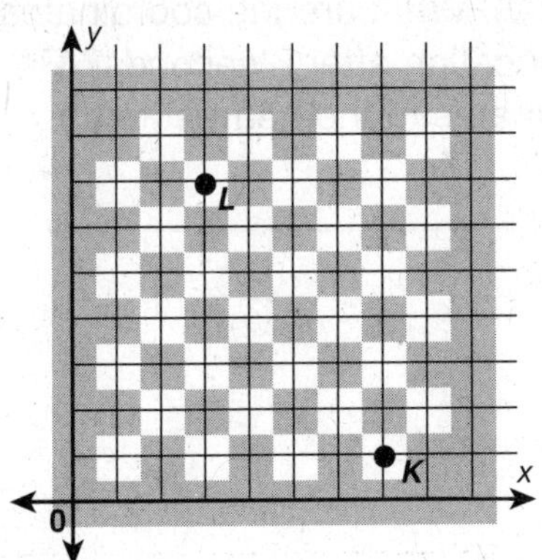

2. The preimage of M' has coordinates $(-6, 5)$. What is the vector that translates $\triangle MNP$ to $\triangle M'N'P'$?

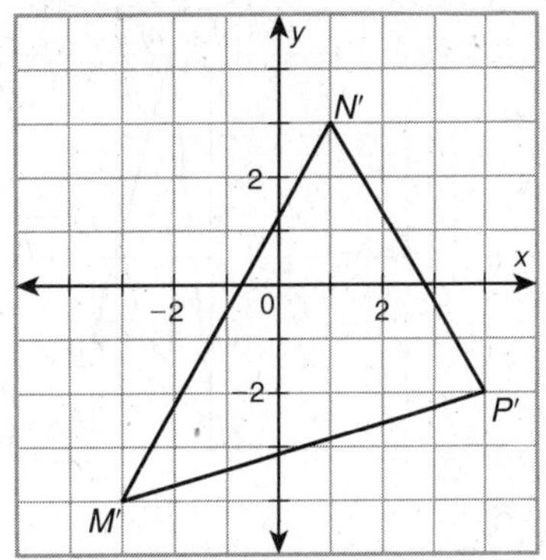

3. In a quilt pattern, a polygon with vertices $(3, -2)$, $(7, -1)$, $(9, -5)$, and $(5, -6)$ is translated repeatedly along the vector $\langle 4, 5 \rangle$. What are the coordinates of the third polygon in the pattern?

4. A group of hikers walks 2 miles east and then 1 mile north. After taking a break, they then hike 4 miles east and set up camp. What vector describes their hike from their starting position to their camp? Let 1 unit represent 1 mile.

Choose the best answer.

5. In a video game, a character at $(8, 3)$ moves three times, as described by the translations shown at right. What is the final position of the character after the three moves?

Move 1: $\langle 2, 7 \rangle$
Move 2: $\langle -10, -4 \rangle$
Move 3: $\langle 1, -5 \rangle$

A $(-8, 3)$ **C** $(1, 1)$

B $(-7, -2)$ **D** $(9, 2)$

6. The logo is translated along the vector $\langle 8, 15 \rangle$. What are the coordinates of R'?

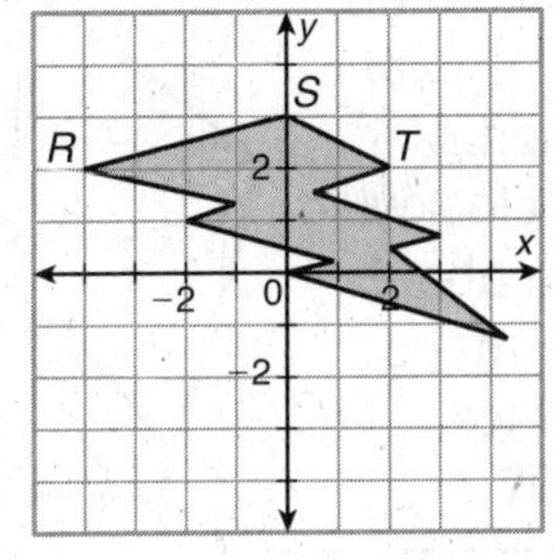

F $(4, 17)$ **H** $(15, 18)$

G $(12, 17)$ **J** $(11, 19)$

7. $\triangle DEF$ is translated so that the image of E has coordinates $(0, 3)$. What is the image of F after this translation?

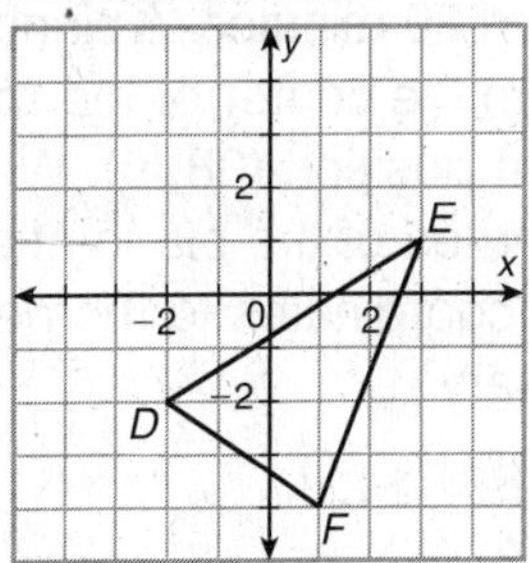

A $(1, -1)$ **C** $(-2, -2)$

B $(4, -2)$ **D** $(-2, 6)$

Holt Geometry

Problem Solving
Rotations

1. $\triangle ABC$ is rotated about the origin so that A' has coordinates $(-1, -5)$. What are the coordinates of B'?

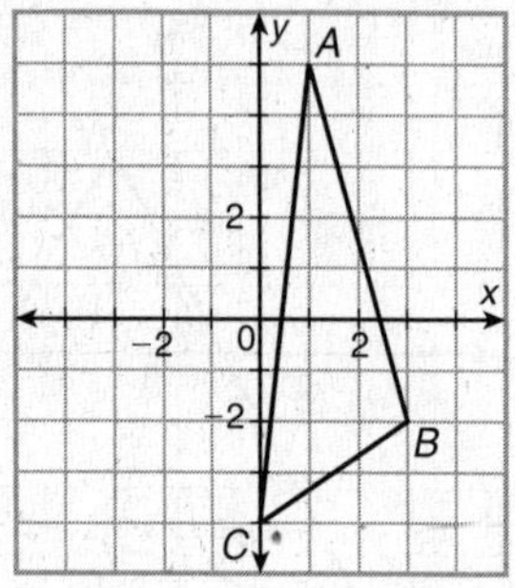

2. A spinning ride at an amusement park is a wheel that has a radius of 21.5 feet and rotates counterclockwise 12 times per minute. A car on the ride starts at position $(21.5, 0)$. What are the coordinates of the car's location after 6 seconds? Round coordinates to the nearest tenth.

____

3. To make a design, Trent rotates the figure 120° about point P, and then rotates that image 120° about point P. Draw the final design.

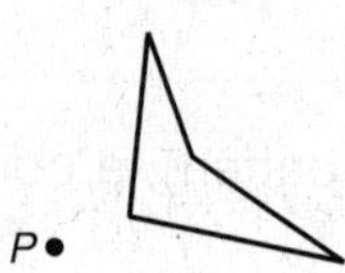

Choose the best answer.

4. Point K has coordinates $(6, 8)$. After a counterclockwise rotation about the origin, the image of point K lies on the y-axis. What are the coordinates of K'?

   A $(0, 5)$ **C** $(0, 8)$

   B $(0, 6)$ **D** $(0, 10)$

5. $\triangle NPQ$ has vertices $N(-6, -4)$, $P(-3, 4)$, and $Q(1, 1)$. If the triangle is rotated 90° counterclockwise about the origin, what are the coordinates of P'?

   F $(-4, -3)$ **H** $(3, 4)$

   G $(-4, 3)$ **J** $(3, -4)$

6. The Top of the World Restaurant in Las Vegas, Nevada, revolves 360° in 1 hour and 20 minutes. A piano that is 38 feet from the center of the restaurant starts at position $(38, 0)$. What are the coordinates of the piano after 15 minutes? Round coordinates to the nearest tenth if necessary.

   A $(0, 38)$

   B $(-38, 0)$

   C $(14.5, 35.1)$

   D $(35.1, 14.5)$

7. The five blades of a ceiling fan form a regular pentagon. Which clockwise rotation about point P maps point B to point D?

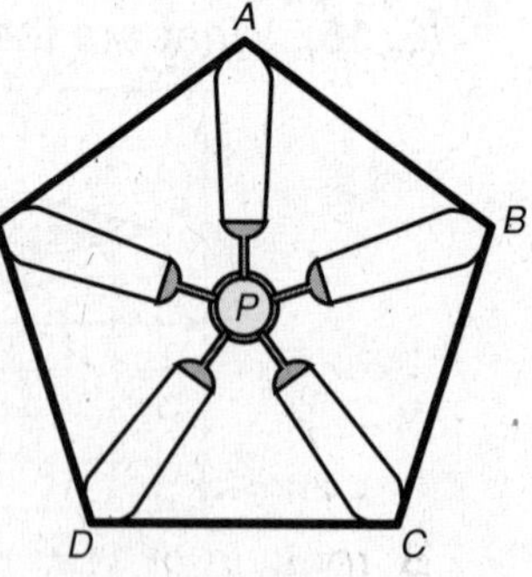

   F 60°   **H** 120°

   G 72°   **J** 144°

Holt Geometry

Problem Solving
Compositions of Transformations

1. A pattern for a new fabric is made by rotating the figure 90° counterclockwise about the origin and then translating along the vector $\langle -1, 2 \rangle$. Draw the resulting figure in the pattern.

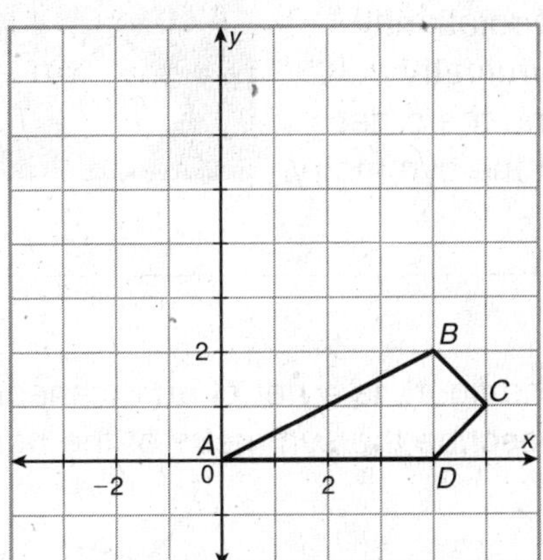

2. $\triangle LMN$ is reflected across the line $y = x$ and then reflected across the y-axis. What are the coordinates of the final image of $\triangle LMN$?

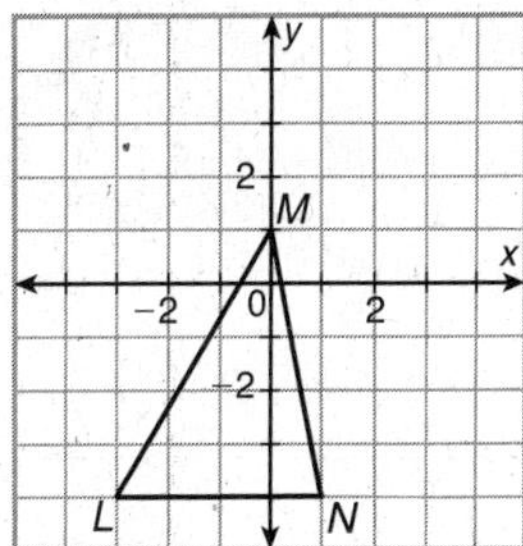

____

Choose the best answer.

3. $\triangle EFG$ has vertices $E(1, 5)$, $F(0, -3)$, and $G(-1, 2)$. $\triangle EFG$ is translated along the vector $\langle 7, 1 \rangle$, and the image is reflected across the x-axis. What are the coordinates of the final image of G?

  A $(6, -3)$ **C** $(-6, 3)$

  B $(6, 3)$ **D** $(-6, -3)$

4. $\triangle KLM$ with vertices $K(8, -1)$, $L(-1, -4)$, and $M(2, 3)$ is rotated 180° about the origin. The image is then translated. The final image of K has coordinates $(-2, -3)$. What is the translation vector?

  F $\langle 6, 4 \rangle$ **H** $\langle -1, -11 \rangle$

  G $\langle 6, -4 \rangle$ **J** $\langle -10, -2 \rangle$

5. To create a logo for new sweatshirts, a designer reflects the letter T across line h. That image is then reflected across line j. Describe a single transformation that moves the figure from its starting position to its final position.

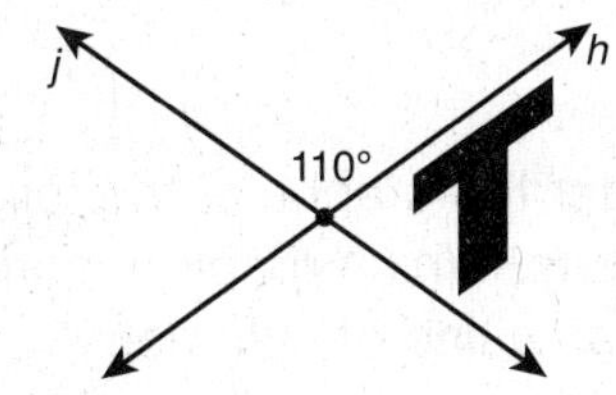

  A translation

  B rotation of 110°

  C rotation of 220°

  D reflection across vertical line

6. Which composition of transformations maps $\triangle QRS$ into Quadrant III?

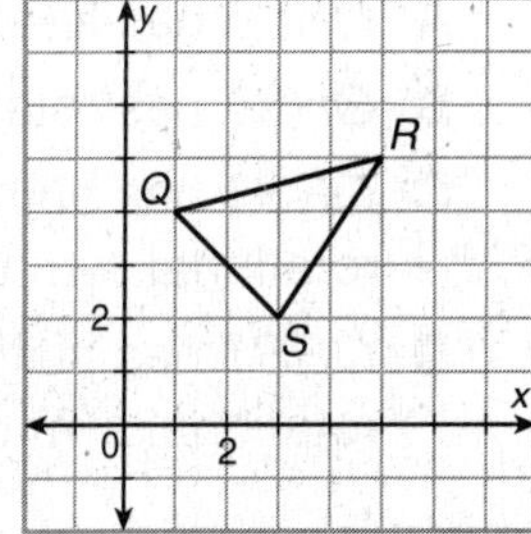

  F Translate along the vector $\langle -6, 4 \rangle$ and then reflect across the y-axis.

  G Rotate by 90° about the origin and then reflect across the x-axis.

  H Reflect across the y-axis and then rotate by 180° about the origin.

  J Translate along the vector $\langle 1, 2 \rangle$ and then rotate 90° about the origin.

Holt Geometry

Problem Solving
Symmetry

1. Tell whether the window has line symmetry. If so, draw all the lines of symmetry.

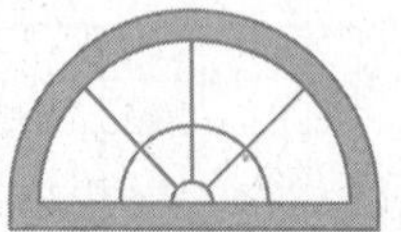

2. Tell whether the quilt block design has rotational symmetry. If so, give the angle of rotational symmetry and the order of the symmetry.

3. Tell whether the hemisphere has plane symmetry, symmetry about an axis, both, or neither.

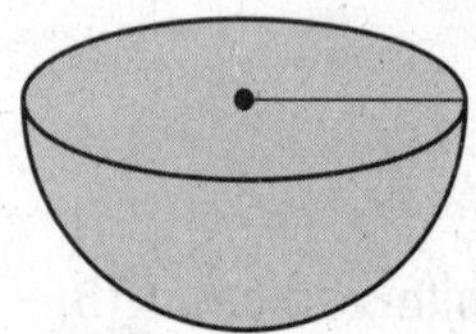

4. The figure is a net of an octahedron. Describe the symmetry of the net.

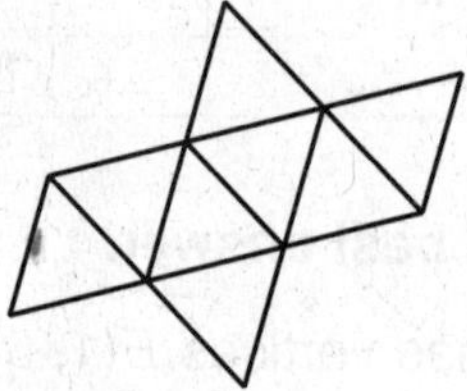

Choose the best answer.

5. Which is a true statement about the figure with vertices $Q(-2, -4)$, $R(0, 1)$, $S(8, 1)$, and $T(5, -4)$?

  A $QRST$ has line symmetry only.

  B $QRST$ has rotational symmetry only.

  C $QRST$ has both line symmetry and rotational symmetry.

  D $QRST$ has neither line symmetry nor rotational symmetry.

6. What is the order of rotational symmetry for the figure shown?

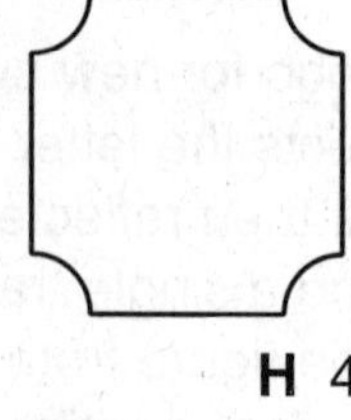

  F 2   **H** 4

  G 3   **J** 6

7. Which of these figures has exactly three lines of symmetry?

  A 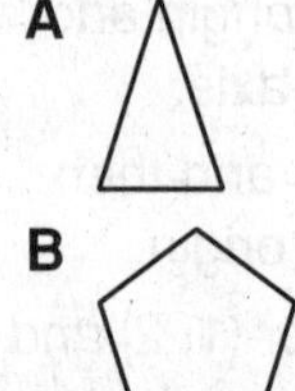   **C**

  B 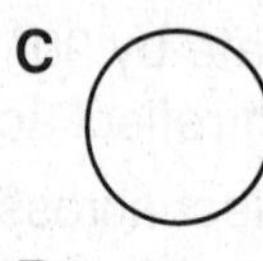   **D**

8. Consider the graphs of the following equations. Which graph has the line $x = 3$ as a line of symmetry?

  F $y = x^2 + 3$

  G $y = (x + 3)^2$

  H $y = (x - 3)^2$

  J $y = x^3$

Holt Geometry

<table>
<tr><td>LESSON 12-6</td><td></td></tr>
</table>

Problem Solving
Tessellations

1. Identify all the types of symmetry (translation, reflection, or rotation) in the pattern.

2. Mara made the beaded bracelet shown below. Identify the symmetry in the main pattern of the bracelet.

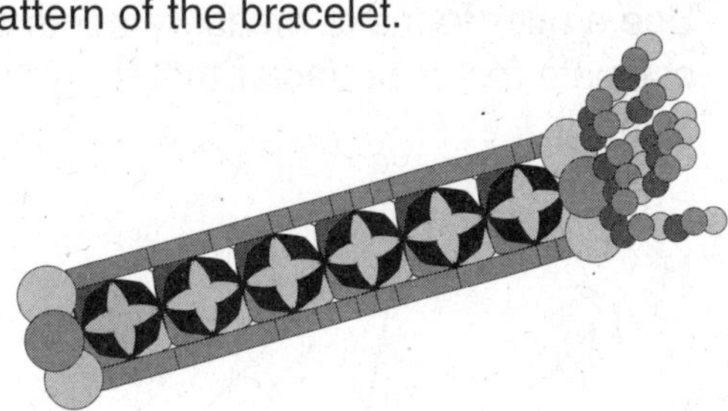

Choose the best answer.

3. Classify the tessellation.

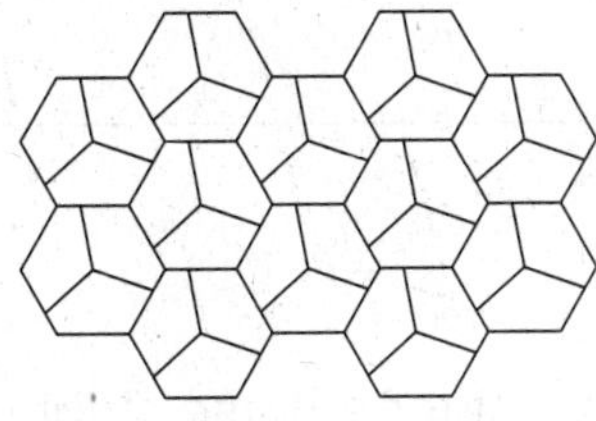

A regular

B semiregular

C neither regular nor semiregular

D part regular and part semiregular

4. The square tile below is formed from a tessellation of polygons. Which of the following types of symmetry is in the tile?

F rotation   **H** translation

G glide reflection   **J** no symmetry

5. Which is a true statement about the pattern?

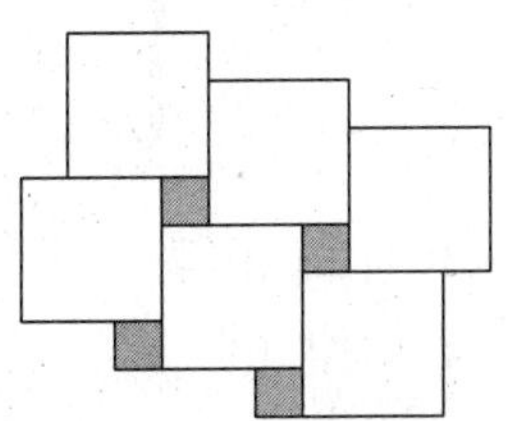

A It is a regular tessellation because it is made from squares.

B It is a semiregular tessellation because the squares are not congruent.

C It is a tessellation that is neither regular nor semiregular.

D The pattern does not form a tessellation.

6. Which is a true statement about the parallelograms shown below?

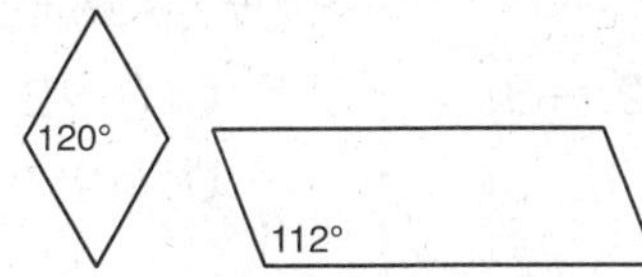

F They can be used to form a semiregular tessellation.

G They can be used to form a tessellation that is neither regular nor semiregular.

H They cannot be used to form a tessellation.

J It is impossible to tell whether the figures can form a tessellation.

Holt Geometry

Problem Solving
Dilations

1. An artist is designing wallpaper by dilating triangles such that $\triangle KLM \rightarrow \triangle K'L'M'$. Use a ruler to make measurements and estimate the scale factor that the artist is using.

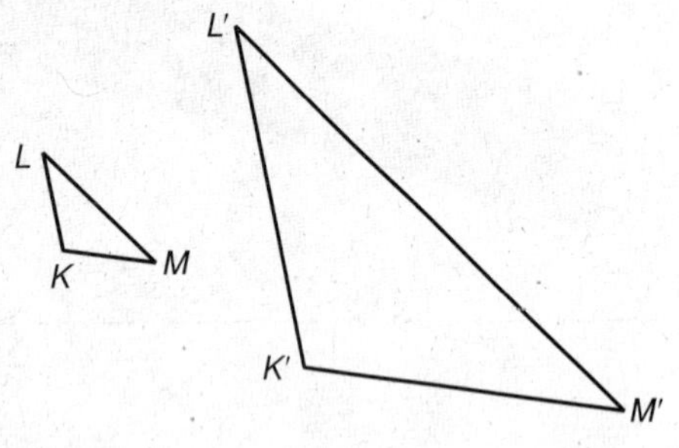

2. $\triangle DEF$ is transformed by a dilation centered at the origin. What scale factor produces an image that has a vertex at $D'(-1, -1)$? Find the coordinates of the other two vertices after the dilation.

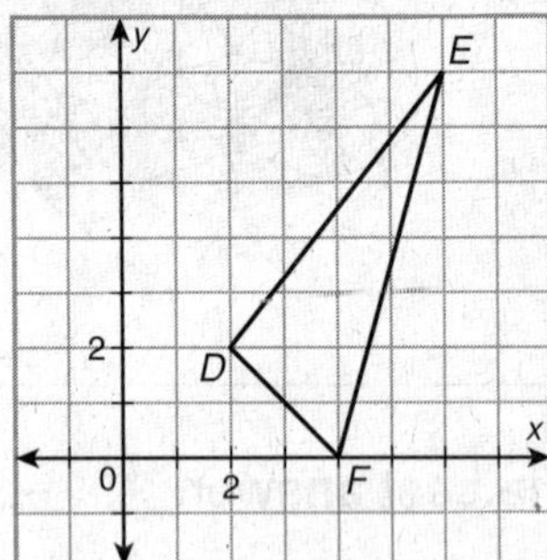

Choose the best answer.

3. $\triangle STU$ is dilated with a scale factor centered at the origin so that T' has coordinates $(-9, 6)$. What are the coordinates of S'?

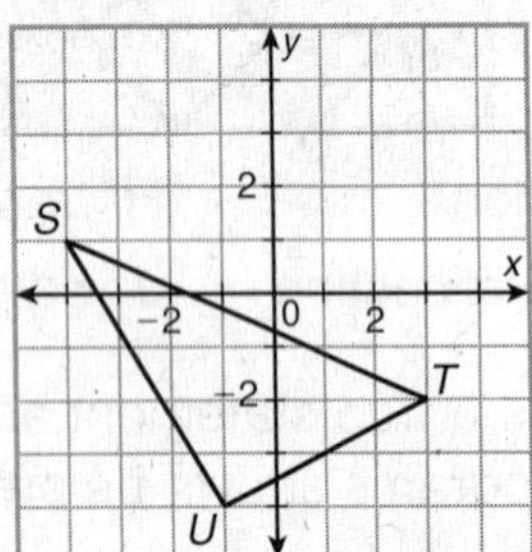

A $(12, -3)$ **C** $(-12, 8)$

B $(-12, 3)$ **D** $(12, -5)$

4. A blueprint for a horse stable shows a reduction of the stable using a scale factor of $\frac{1}{24}$. In the blueprint, a horse stall is shown by the diagram below. What is the actual area of the stall?

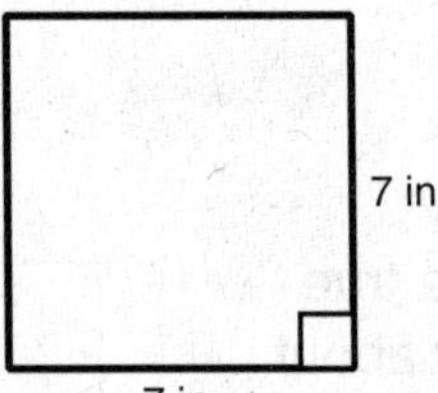

F 144 ft^2 **H** 576 ft^2

G 196 ft^2 **J** 1176 ft^2

5. Steven is enlarging a photograph by a scale factor of 2.5 and then placing 2-inch matting around the perimeter of the enlarged photograph. If the photograph is 3 inches by 5 inches, what will be the area of the matting?

A 37.5 in^2 **C** 96 in^2

B 93.75 in^2 **D** 189.75 in^2

6. What is the scale factor of a dilation centered at the origin that maps $A(5, -6)$ to $A'(-15.5, 18.6)$?

F 2

G 3

H -2.1

J -3.1

Holt Geometry